The Birds of Babel

Satellites for a Human World

Hal Glatzer is a journalist and television producer, specializing in news of the information industries. He describes himself as an "explainer," helping nontechnical people understand new technology and helping specialists in computers and telecommunications to understand one another. Mr. Glatzer is a regular contributor to magazines for computer users and to trade journals for information industry executives. An accomplished guitar player, he also writes for music magazines.

Born in New York, Mr. Glatzer earned a BA in English at Syracuse University, then moved to Hawaii where he pursued a career in journalism—both in print and on television—and earned a Master's degree in Communication at the University of Hawaii. He now lives in Seattle, Washington.

The Birds of Babel

Satellites for a Human World

by Hal Glatzer

Howard W. Sams & Co., Inc.
4300 WEST 62ND ST. INDIANAPOLIS, INDIANA 46268 USA

Indianapolis, IN 46268

FIRST EDITION
FIRST PRINTING—1983

International Standard Book Number: 0-672-22033-4
Library of Congress Catalog Card Number: 83-50167

Edited by: Jim Rounds
Illustrated by: R. E. Lund

Printed in the United States of America.

PREFACE

Five years ago, if someone tried to tell you that soon you'd need a computer for your small business, what would you have done? Would you have laughed? Would you have said, "Come on! Nobody needs computers except a bank, or an insurance company. Besides, computers are too big and expensive for the average person to have one." Or would you have bought a small, personal-size computer, as some people did, learned what it could do, and made it a part of your life? Today, many of these people who saw the opportunities in small computers are working in a practically recession-proof industry: making computers, programming them, and selling that hardware and software to the rest of us.

If you missed the first wave—the computer wave—don't despair. There are fortunes yet to be made, and there are intangible rewards waiting for people who help people to understand one another. The next wave of technology is bringing both opportunities. It is the wave called *telecommunication.*

When I told people I was writing a book about satellites, many of them looked at me and said, "Come on! Nobody needs a satellite except a tv network or the phone company. Besides, satellites are too big and expensive for the average person to have one."

This time, the objection is a little closer to the mark than the computer put-down was. Satellites *are* very expensive, and the average person cannot simply go out and buy one as he or she can buy a microcomputer. But that's not the only way to deal with satellites. If you have a television or a telephone, you are *already* a satellite user. Virtually everyone in the industrialized countries is a user of satellites in their business, in their recreation, and in their personal communication. The leaders of the less-developed countries know that having access to a satellite can help supply their people with sustenance instead of starvation; provide education instead of ignorance; and establish nationhood instead of anarchy.

Telecommunication—especially with satellites—is a bridge to new ways of life. The people who understand it will be among the most valuable and sought-after professionals for the rest of their lives. Their spheres of influence will grow immensely. Those with *original* ideas to contribute will find themselves at the very crest of the wave.

This book explains *in plain language* how satellites work. But, more importantly than that, it shows how satellites are changing our lives. This book was written for people who care about what happens to them, and about what happens to other people, whether they live next door or far away. Telecommunication has brought the lives of other people into our own. When we agree to use it, or control it, we take on great responsibilities. When a person uses a small computer, there isn't much *communication* going on: a computer, even one that is "user-friendly," is still a machine. But when a person uses a satellite, there is some person-to-person communication going on, and that is what makes us human.

This book will help nontechnical people understand the new technologies, and it will also help specialists in one field understand what colleagues in other fields are doing. But it will also demonstrate that telecommunications is more than technology alone. True communication comes from an understanding of what the other person thinks and feels. I have written this book—and its companion volume, *Who Owns the Rainbow?*—with *that* kind of communication in mind.

Satellite technology is not the only thing of importance, but it affects all the others. Therefore, the first three chapters of this book explain—in simple language—how satellites work. Great inventions almost always turn natural phenomena into practical applications. For communication satellites, the most important natural phenomenon is the orbit in which they ride, and it gets a chapter of its own (Chapter 2: The Geostationary Orbit). Human endeavors also require that trade-offs be made, and balances struck, between competing needs. Those are explained in Chapter 3 (Trade-Offs in Satellite Communication).

The proliferation of satellites has solved some problems and created new ones. Because communication has been lumped with transportation for practically all of human history, it is helpful to look at the origins of our communications laws (Chapter 4: Parallels of History). International and domestic legal issues are far from settled, as Chapters 5 (Space Law) and 6 (Piracy on the High Frontier) show.

Satellites have brought political and social changes to the countries that developed them; Chapters 7 (INTELSAT) and 8 (INMARSAT) put them in perspective. The "third world" of developing countries has been even more profoundly affected; Chapters 9 (ATS-1): Promises and Compromises) and 10 (Satellites for the Pacific) present those changes in human terms.

The major providers of—and customers for—worldwide communication are, as always, very large organizations. Until the 20th century, only governments had the resources to build or maintain such networks; now, those networks are increasingly guided by private and transnational corporations. Chapter 11 (Very High Finance) explains who really pays for satellite communication. The financial considerations are covered in Chapter 12 (Financing Satellites). Satellites now have the potential to be bought and sold as real estate: how and why are examined in Chapter 13 (Transponders for Sale: Real Property in Outer Space).

The next "generation" of satellites will bring television and other programming right into people's homes, as Chapter 14 (Direct-to-Home Broadcast Satellites [DBS]) explains. But there are other kinds of satellites, too: those that take pictures from space, Chapter 15 (Pictures From Space), and those that help search for lost or missing people on earth, Chapter 16 (Position-Locating Satellites). And—lest anyone forget the competition—satellites are not the only new technology around. Chapter 17 (Don't Count Cables Out) warns us: satellites aren't the only answer.

This book was written to help put a feeling for *people* into what is often seen as a business dominated by insensitive technocrats. The final chapter, Chapter 18 (We Are Silent), shows what can happen when technology moves on, and the people are left behind.

As Albert Einstein said, "It is not enough that you should understand about applied science in order that your work may increase man's blessings. Concern for man himself and his fate must always form the chief interest of all technical endeavors . . . in order that the creations of our mind shall be a blessing and not a curse to mankind. Never forget this in the midst of your diagrams and equations."

About the Cover

One of the first satellites to be launched was called "Early Bird." Thus the nickname "birds" was applied to all satellites shortly thereafter. The cover is a symbolic interpretation of satellites or "birds" and their role in the improvement and expansion of world communications. Each element of the design is not intended to be taken literally. Rather, it serves to convey the interaction of all aspects—dishes, satellites, and countries—which pertain to global communications today. The goal is to suggest the social and political importance satellites have all over the world.

One of the great books about communication is Erik Barnow's *A Tower in Babel*; the first of a three-volume history of broadcasting. From that book—and not the Bible—is the title of this book derived.

CONTENTS

Chapter 12

Chapter 13

Section 5. New Opportunities

Chapter 14

Chapter 15

Chapter 16

Chapter 17

Chapter 18

ACKNOWLEDGMENTS

The idea for this book, and for *Who Owns the Rainbow?,* came from meeting several remarkable and farsighted people in Hawaii in 1977. Richard Barber, John Bystrom, and Stan Harms, more than anyone, gave me a feeling for how international telecommunication could be both technically elegant and humanitarian.

At about the same time, Norm Abramson, Karen Ah Mai, Jim Dator, Frank Derfler, Herbert Dordick, Stan Harter, Paul Heinberg, Nicholas Johnson, Meheroo Jussawalla, Sarah Sanderson King, Libert Landgraf, Ted Merrill, Carol Misko, Syed Rahim, Jim Richstad, Marcellus Snow, John Southworth, Robert Theobold, Don Topping, Vijay Ram Trehan, Robert Walp, and Dan Wedemeyer gave me insights and observations that I might never have found alone.

When the time came to write, I was able to draw on the expertise of many valuable resource people, especially Veronica Ahern, Richard Butler, Wilson Dizard, Harvey Levin, Andrew Lipman, Glen Pafumi, Joseph Pelton, Edward Ploman, Jon Post, Edward Reinhart, Delbert Smith, David Williams, and Robert Wold.

My deepest thanks go to Jonathan Miller and Will Pape, who helped me shape what I had written into a book.

Section 1

UNDERSTANDING THE TECHNOLOGY

"A bird of the air shall carry the voice.
. . . and that which hath wings
shall tell the matter."
–Ecclesiastes

Chapter 1

AFTER COMPUTERS ...COMMUNICATION

Computers are hot. They clamor for public attention. They grow faster and more useful yet, at the same time, smaller and cheaper. It's a paradox so challenging that it's captured the headlines. The public knows and cares about computers—and the information media are fixated on them—in large part because they are so *real.* When that box full of hardware and software plays games, it's fun; when it's working as a cost accountant, it's more important than a ledger. Computers are now touching our lives in many exciting ways. Sales climb to new heights every year. Computer "literacy" is now expected of even the very youngest children. *Time* magazine made the statement that "the computer" was qualified to be 1982's " 'Man' of the Year."

But there is another revolution that has been obscured by the noise and smoke of computers and their advocates. It is a revolution in *communication.*

Most people do not know why the cost of a long-distance telephone call is a bare fraction of what it was 20 years ago; only a handful have seen it as an opportunity to get rich. Most people move money in and out of their banks with the push of a button and care little for the sciences that made it possible; yet those who understand it are likely to be running the financial institutions of the 1980s and '90s. Most people watch television programs from around the world, but only a few understand the importance of the satellites that make it possible.

Those people who *do* learn how to use the new communication tools will be one step ahead of those who do not; just as people who learn to use computers gain a lead over those who do not. Fortunately, in the countries which have embraced computers, the technology has made it possible for practically anyone to learn the fundamentals. The trend has been toward smaller, cheaper machines that are easier and easier to operate.

WHO NEEDS TO KNOW?

It would be flippant to say that *everyone* should understand telecommunication, and satellites in particular. But the fact is, they are involved in so many people's lives already that it is difficult to find a person or an interest group that is not affected.

Farmers, sailors, merchants, and golfers depend on satellites for weather reports. Prospectors get geological pictures, and rescue teams get help finding people in distress.

Tv news reporters and entertainers reach more homes than ever before because their images are carried by satellite. Publishers print newspapers and magazines all over the U.S. and the world with the help of satellites. The armed services have several of their own satellites, but they also depend on civilian satellites that are in orbit to back them up.

Americans once watched a heart bypass operation, live on television, by satellite; doctors, nurses, and hospitals have been doing that for years.

Teleconferencing by satellite lets far-flung people do business "face-to-face," even when they're thousands of miles apart. The U.S. House of Representatives does its work in full view of the nation, by satellite.

Ordinary people, in businesses large and small, use satellites practically every time they make a long-distance call through the phone company or through an alternative service.

The list is not complete, because new uses for satellites are invented almost every day. Satellites are not esoteric; they are practical.

It is possible that, as computers and communication technologies proliferate, a new social schism may emerge which divides the information-rich from the information-poor. But that will not happen if the people who become information-savvy also learn generosity and sensitivity. The potential for kindness is there, but very few of the textbooks, the schools, or the media make a point to stress the *responsibilities* which should accompany the power of computers and communications.

BEYOND TECHNOLOGY IS HUMANITY

There is much more to satellites than just the technology for building and operating them. James Martin, whose books on telecommunications and computer communications are among the principal reference works in this enormous field, distinguishes the motivations of two opposing "teams" in the satellite game. He suggests that users—whether they are large organizations or small and occasional customers—are set against the operators of satellites because their respective needs, and the right technical responses to meet those needs, are not the same.

For example, where independents want to use satellite channels at the lowest possible cost, the operators want guaranteed rates of return; where independents want access to the satellite from many different places, operators want to concentrate traffic into a few, high-volume locations; where independents want to have access available at any time, on demand, operators want to fill up their circuits as soon as possible after a satellite is launched.

If these discrepancies concerned only electronics engineers and system managers, this book might never have been written—or even been needed. But those choices involve the nontechnical, nonspecialist users directly. They need to exert influence on the operators, if they want satellites built to accommodate their needs. To do so, they will need to learn what is possible and what is not. Users and consumers of satellite services *can* have a great deal to say about the way those satellites are designed and built. Political, social, and economic disagreements are raging around the world today because, for too long, users and operators did not wish to understand each other.

Most of the rest of the world has little or none of the information resources which the U.S. and the other industrial nations take for granted. But they are painfully aware that such resources exist, and that—with them—they might lift their countries out of poverty. That is no armchair speculation: doctors with access to a rural telephone network are better able to fight epidemics than those without it; a government that can produce its own television broadcasts is more likely to engender literacy than one that relies on imported programs presented in a foreign language.

People of the developed countries need to understand the technology of communication, but that alone is not enough. There are serious political, social, legal, and economic problems which surround communication and which—for all the progress of technology—stubbornly refuse to go away. People who understand communications can use their knowledge to rise in big organizations, or to start their own. They can become leaders in their fields: science, law, medicine, finance, diplomacy, sales, teaching . . . anything that is based on the exchange of information. They will be rewarded for their efforts not only with money but with the thanks of the world *if they also help to dispel the frustration, resentment, and anger that challenges them.*

WHERE DO WE START?

Ideas are intangible. Satellites are real. After radios, they are probably the most important communication tools of the twentieth century. The importance of satellites is just now being felt, yet satellites remain something of a mystery even to people whose work or leisure depends on them. Why? For one thing, satellites are invisible. When a person makes a phone call, he or she doesn't really care whether the conversation will be transmitted by cable, microwave, or satellite, as long as it gets through. For another, the cost of operating a satellite is spread over so many customers that none of them seems to have any property rights in it. Then, too, the process of designing, launching and maintaining a satellite is far beyond the resources of all but the largest corporations or public institutions; so people feel humbled by the enterprise and probably awestruck that anyone can understand it.

This book was written out of the conviction that practically everyone can *and should* understand satellites for many of the same reasons that people should understand computers: without them, commerce in the 1980s will be nearly impossible. That includes commerce in ideas as well as in goods.

The advantages of using satellites for communication do not leap out at a person in the way that the advantages of using computers for accounting or word processing do. However, the more a person knows about satellites the better he or she will be able to negotiate for their use, and to take advantage of new opportunities as they arise. In just one area alone–direct satellite television broadcasts to homes and offices–there is the potential for new fortunes to be made, and for spinoff industries to grow up supplying both the equipment and the programs. The educational and social opportunities are at least as great as the financial ones.

New kid on the block–Will this sight be common in the 1980s as television antennas have been since the 1950s? (Courtesy COMSAT)

THE BIRDS

On October 4, 1957, the Soviet Union astoundeo the world by launching the first orbiting satellite. The event also launched many young American scientists and engineers on their future careers, as schools tried to teach physics, math, and electronics on an unprecedented scale. Sputnik jolted the American space program because throughout the 1950s, the Air Force had made it seem as though their own satellite ("the size of a basketball," it liked to say) was practically aloft. Even after the Americans did launch Vanguard (January 31, 1958) the Soviets kept one jump ahead for years. A dog named Laika circled the earth in Sputnik II, a vehicle which was *300 times heavier than Vanguard!* Within four years, the Soviet cosmonaut Yuri Gagarin waved hello to the earth from space. This angered and frustrated the Americans, because cold-war logic held that the U.S.S.R. was a backward, repressive nation unable even to

ESTIMATED EARTH STATION SALES

Year	Main System Stations	Medium Density	Thin Route	TVRO Cable	TVRO Hotel	TVRO Home	Broadcast Transmission	DBS	Private Network	Scientific	Military	Total
1981	$10	$ 90	$ 3	$30	$ 1	$ 30	$1	—	$ 48	$ 3	$216	$ 432
1982	10	120	2	30	4	50	1	—	56	3	220	496
1983	15	120	3	30	8	100	1	—	120	3	225	625
1984	15	120	3	23	8	100	2	—	120	4	230	625
1985	15	120	5	23	8	200	2	—	130	5	235	743
1986	20	120	13	23	8	200	3	$100	140	6	240	873
1987	20	90	25	23	12	300	3	150	150	8	245	1,026
1988	20	75	50	30	12	300	3	200	160	10	250	1,110
1989	25	75	100	30	16	400	4	250	170	13	255	1,338
1990	25	60	150	30	16	500	4	300	180	15	260	1,540
1991	25	60	175	30	16	600	4	500	190	18	265	1,883

NOTE: ALL DOLLAR AMOUNTS ARE IN MILLIONS.

Source: Frost & Sullivan

Estimated Earth Station Sales—A research firm estimates that the sales of satellite earth stations may reach nearly $1.9 billion by 1992—*a fourfold increase from the baseline over ten years. Not all segments of the market will grow so much, however: receive-only television (ROTV) dishes for cable tv operators may have already reached their peak; military procurement may show only a slight increase. But direct-to-home (DBS) antenna sales could quintuple once they begin; private networks, such as those owned by large corporations, could triple or quadruple their dish purchases; thin-route antenna sales (e.g., for rural or third-world operations) could increase 50-fold.*

---

feed itself. That may very well have been true, but it was also a nation whose people were able to make great sacrifices and expend enormous energy in their national interest.

The Americans' ace-in-the-whole, however, was Bell Laboratories, the heir to a home-grown tradition of inventing and experimenting that had been embodied in its founder, Alexander Graham Bell. (Actually, the well-financed, industrial, state-of-the-art laboratory for making inventions was the legacy of Thomas Alva Edison.) Under contract to the armed services, Bell Labs built the first American communication satellites.

One of the first to be launched was called "Early Bird" and it was like a parrot, in the sense that it reflected back whatever signals were beamed up at it. The nickname "birds" got stuck on satellites shortly thereafter and, like many nicknames, there is some justification for using it. Birds and satellites both "fly"; they both call or sing in many different ways; they have been domesticated, and yet they still generate a sense of wonder and mystery. Thinking about both has encouraged human beings to seek things that are far away, to travel, and to experience a kind of freedom.

One of the great books about communication is Erik Barnow's *A*

INTELSAT I
(Early Bird)
240 circuits or
1 TV channel

INTELSAT II
240 circuits or
1 TV channel

INTELSAT III
1500 circuits,
up to 4 TV
channels

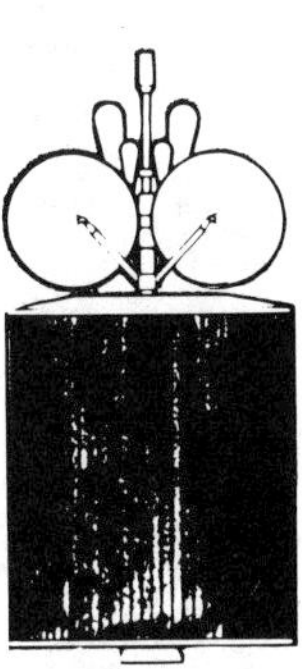

INTELSAT IV
3750 circuits
plus 2TV channels

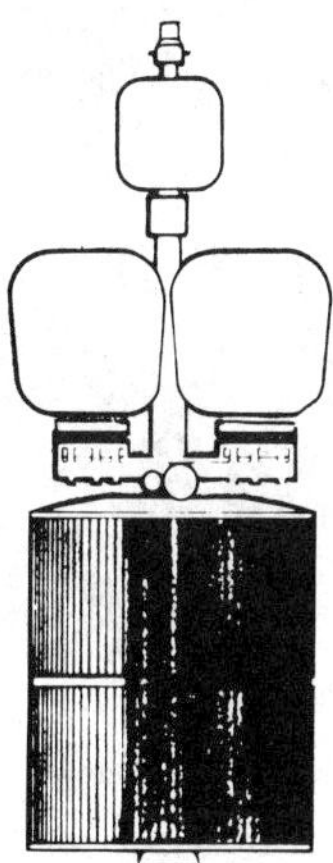

INTELSAT IV-A
6250 circuits
plus 2 TV channels

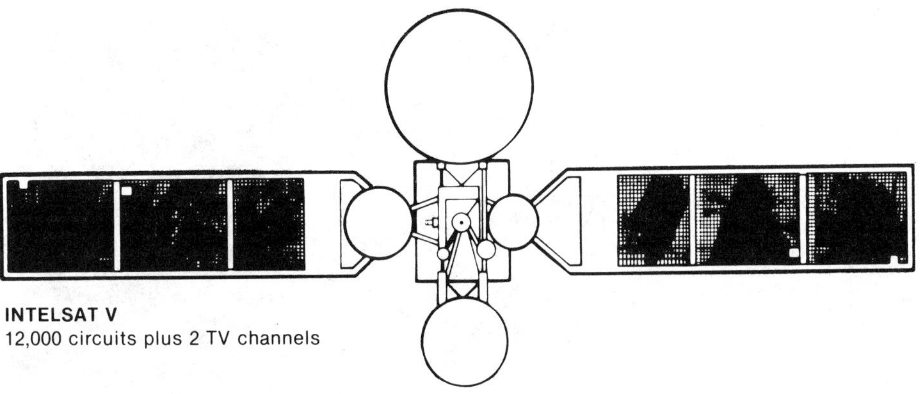

INTELSAT V
12,000 circuits plus 2 TV channels

Evolution of the Birds—*The following series of photographs shows the evolution of INTELSAT's satellites and their growth in size. Their circuit capacities have also increased proportionally. The illustration beside each photograph shows the six satellites drawn to comparative scale.*

INTELSAT I
(Early Bird)

INTELSAT II

Early Bird—INTELSAT I—was scarcely larger than a child. INTELSAT II was about as big as two grown men; INTELSAT III was slightly larger. The outer surface of the "barrel" is covered with solar panels to generate electricity from the Sun.

INTELSAT III

INTELSAT IV

With the development of INTELSAT IV and IVA, the satellites had grown considerably to three and four times the height of a man.

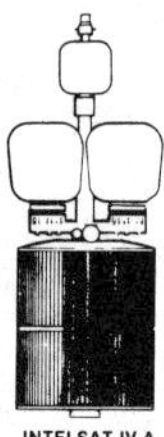

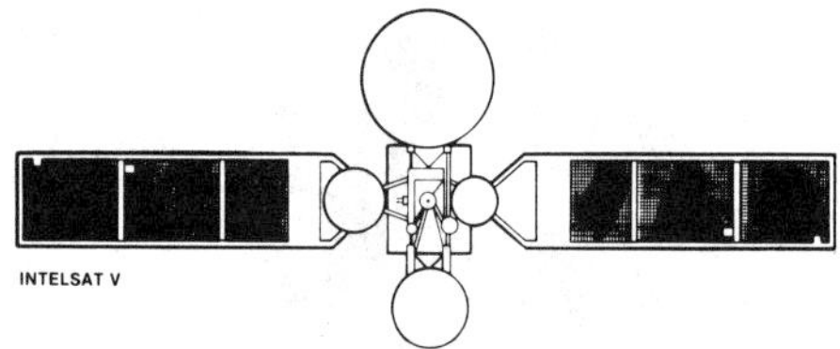

INTELSAT V uses a different satellite design, one with "wings" of solar panels. But the antenna array with its support circuitry is nonetheless much larger than the technician working below it on the laboratory floor. (Photographs: Courtesy COMSAT; Drawings: Courtesy INTELSAT)

Tower in Babel; the first of a three-volume history of broadcasting, it deals with the rise of radio. From that book—and not from the Bible—is the title of this book derived; there are no birds in the short story of Babel. But the birds which we call satellites *are* our new towers in Babel. They soar far higher than any skyscraper and, like carrier pigeons, they bring us news from the far corners of the known world. They can help us to understand people who speak entirely different languages, and in so doing, they can help us to understand ourselves.

THE IMPORTANCE OF THE BIG PICTURE

New technology tends to ride the coattails of its predecessors for a while before coming into its own. We laugh, today, at the expressions "iron horse," "horseless carriage," "moving picture," and "wireless telegraph" because they are quaint. But what they did, at the time, psychologically, was to place a new—possibly threatening—idea in a familiar context. Because most people seem to feel comfortable with that process, their public institutions follow suit. Instead of creating entirely new organizations, they expand existing ones to accommodate fresh responsibilities. The limitation, of course, is that the new idea may not, *in fact,* be an extension of the old one; however comfortable people may be with its colloquial expression.

A famous story illustrates this mental block. A century ago, the chief engineer of the British Post Office pooh-poohed the American invention of the telephone. He is alleged to have said that, while the Americans needed it to cover their vast territory, the British could do without it because "we have so many messenger boys." He had missed the importance of the new telephone, as a communication medium in its own right, because he thought of it merely as an extension of the telegraph with which he was familiar.

Chapter 2

THE GEOSTATIONARY ORBIT

THE CROWDED TOWERS OF BABEL

Hilltops and other view properties are desirable places to live. But once an apartment house occupies a choice view lot, it is not possible to build another building on the same site. With flexible zoning regulations and a creative architect, a developer can maximize the number of new apartments without interfering with other residents' views but—ultimately—the number of units is limited by the size and shape of the site. City planners and zoners who set those limits understand that the hilltop has room for only so many view apartments: sooner or later the hilltop will be "full."

But height is a great asset in other ways. Mountain climbers, window-washers, and tourists on the observation deck of a skyscraper understand why: the higher a person stands, the further away the horizon appears, so more territory can be surveyed. If a living room tv can't get a good picture with "rabbit ears" on top of the set, the householder can mount an antenna on the roof. The idea is that, by being higher, an antenna will pick up more of the broadcast signal.

That's why the highest hills, buildings, or towers of a city usually bristle with radio and tv transmitters: they can reach more receivers from high up than they could closer to the ground. Having more *potential* listeners or viewers usually means that a station will, in fact, draw a wider audience, and that is an advantage in competitive markets such as those in the U.S. Even where there is no competition, as in countries where the government owns or operates the radio and tv, a higher transmitter can reach more people without using more *power,* and that is both a cost-saving feature and an improvement in efficiency.

The practical limit to the height of freestanding antennas was reached in the late 1950s, about the same time the first satellites were launched: a tv tower in Oklahoma stands a little higher than New York's Empire State Building. Newer buildings—the World Trade Center in New York and the John Hancock and Sears towers in Chicago—are taller, but not by much. Each of them is festooned with antennas, but none rises higher than 1800 feet.

Still, everyone who broadcasts wants his or her antenna to be one of the tallest. Competition for choice locations can be fierce, especially in the largest cities where the best, highest places are already occupied. Where no new transmitter can be built without interfering with others, local authorities will probably deny an applicant the right to put one up in that location. It is possible, with sophisticated technology, for several broadcasters to share a single antenna mast, so a mixture of different signals can be transmitted simultaneously without interference. But the height of the tower ultimately determines the range of the transmitters mounted on it and, hence, the area which those antennas can serve.

FLYING BY STANDING STILL

Shortly after World War II, a British wireless operator, a technocrat with a literary bent, became fascinated with rockets. The war had cast a pall on rocketry, but the notion was accepted, among many scientists, that rockets could hurl artificial satellites into orbit around the earth. Arthur C. Clarke gave the idea a novel twist.

An orbiting satellite has to keep moving or it will succumb to gravity. A bullet always falls to earth because it carries no propellant to accelerate it after it is fired. But the force of gravity diminishes as the object gets further away from the earth. At about 22,300 miles above the equator, Clarke deduced, the satellite would need to travel only as fast as the earth itself rotated on its axis: 6879 miles per hour. If it did that, it would not need additional acceleration to stay aloft. To an observer on earth, it would appear to hang, motionless, over one spot. This was called the geostationary phenomenon.

Clarke realized that, from so high up, the visible horizon would be enormous, wider than any that could be seen from a ground-rooted tower. He calculated the area to be nearly one-third the surface of the earth. Such a satellite would be only a scientific curiosity unless it could be put to some practical use. Then the idea of a radio antenna took hold of him. Three satellites, equally spaced around the equator, could cover the whole earth (except for the poles) with radio signals. Ground-based broadcasters could use them as "mirrors" in space. There was already a precedent for it: radio relay and repeater antennas[1]* on hilltops, towers, and tall buildings. What Clarke conceived was a dramatic improvement in worldwide communication. He published his theory, and within 20 years it was a reality.

THE BIRDS COME TO BABEL

It seemed, for a while, that satellites would solve the dilemma of crowded hilltops. A satellite could relay not just one signal but dozens of signals; electronics engineers predicted—correctly—that hundreds or even thousands of signals could be relayed just about as easily.

Like Saturn's rings, the "geostationary orbit" circles the planet at the equator. The "Clarke Orbit," as some people call it, is ring-shaped but narrow and thin, in the sense that a satellite cannot be more than a few miles away from the ring's theoretical axis. A satellite loses its geostationary property—it won't stand still—if it's too near, too far away, too far north, or too far south.

The distance is so great that it's very hard to visualize: 22,300 miles straight up is almost one-tenth of the way to the moon and, wrapped around into a circle, it is only slightly less than the circum-

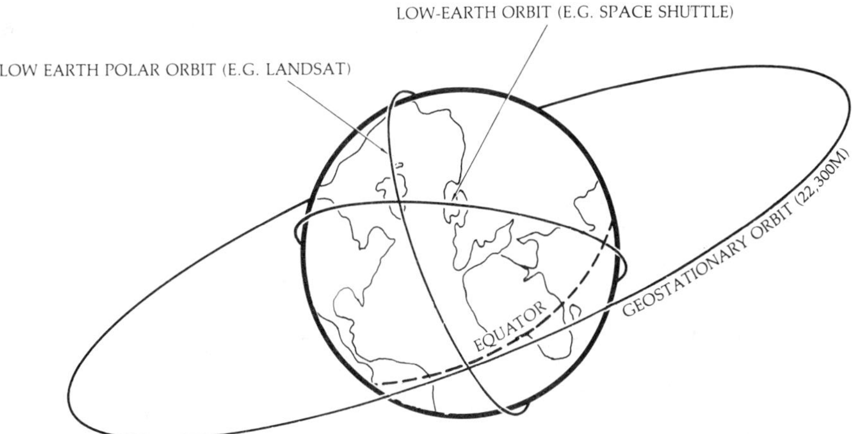

Earth Orbits—Position of an orbit around the Earth depends on the purpose of the satellite or vehicle launched into outer space.

*Reference numbers apply to technical notes placed at the end of each chapter.

ference of the earth at the equator. But the distance of the Clarke Orbit obscures its constricted shape and makes it appear much larger than it really is.

Being a circle, the "geostationary orbit" is seen from the earth as extending from horizon to horizon. A segment of a circle is called an *arc,* so the common name for any piece of the orbit that hangs over a particular place on earth, is the *geostationary arc.* Degrees are used to measure it: each *degree* is 1/360 of the circumference of a circle. At that altitude, one degree equals 720 km (450 mi). Thus, for example, a satellite may be said to be located at 145° West Longitude, with 4° spacing. That means that it is directly over the equator 145° west of the Greenwich meridian, and is separated from any other nearby satellites by at least 4/360 of the whole orbit: 2880 km or 1800 mi.

A satellite is commonly said to be "parked" in a "slot," and the analogy with automobile parking is a very good one. As in a public garage, there are parking places which are more or less desirable for a number of reasons. Those closest to the entry, for example, are preferred over those on the floors farthest away. The best slots fill up first.

The best arc for a satellite to cover the continental United States, for example, is between 54° and 143° W. To cover Alaska and Hawaii as well, from a single satellite, the best arc is between 119° and 143° W. Canada is best served by a narrow arc more or less in the middle of the hemisphere, centered at 100° W. Naturally, any slot which can simultaneously cover the west coast of Europe and the east coast of North America would be ideal for transatlantic traffic, and that is a very desirable slot indeed.

THE LIMITS OF FLIGHT AND FREEDOM

It takes great precision to park a satellite in a geostationary orbital slot. The "drivers" on the ground must monitor it by radio continuously, so that it does not wander away; this is called *telemetry* and the satellite's response is called *stationkeeping.* Telemetry is particularly important while a satellite is being placed in orbit, from the time of its launch to the time that its first relayed signal is received. In 1980, RCA lost contact with a brand new satellite during that critical time; it may be somewhere in orbit or it may have fallen to earth, but without radio control it is useless.

Once in place, a satellite is permitted to roam around the theoretical center of its slot, but on a very short tether. It cannot go

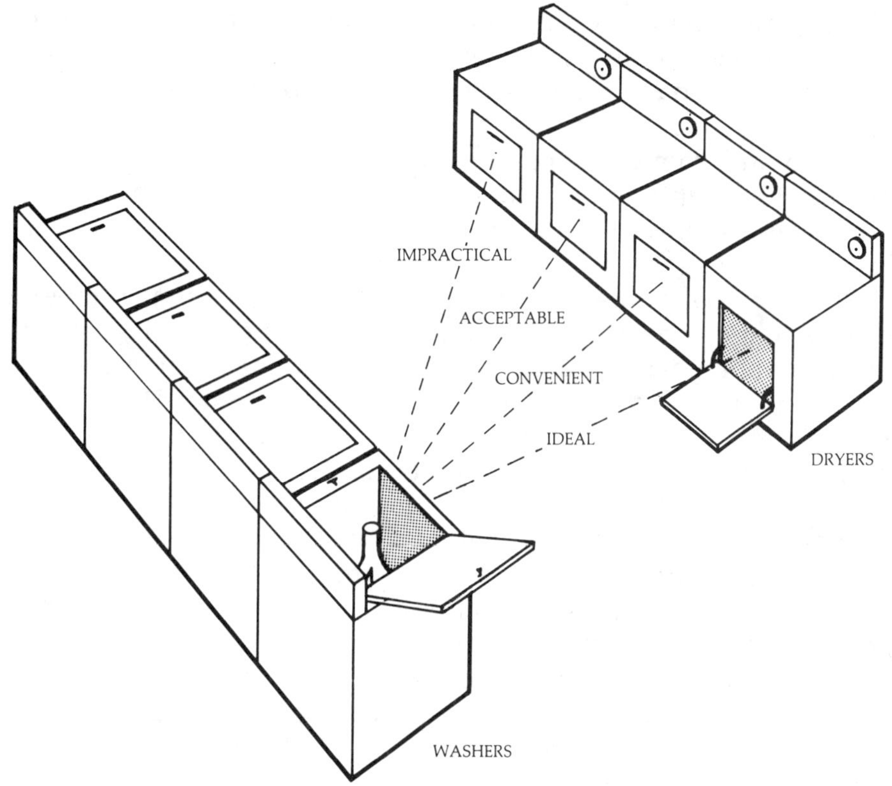

In a public laundromat, the relative position of the dryer with respect to the washer is a measure of desirability.

It is the same with satellites in the orbital arc, with respect to the location of earth stations.

more than about 1/10 of one degree (72 km/45 mi) in any direction. It must never get too close to another satellite. Just as radio transmitters that use the same frequency cannot be located within range of each other, satellites that use the same frequency cannot get too close either. If they did, the signals sent down to earth would interfere with one another, and signals sent up that were intended for one of them might be accidentally received by the wrong one. That could be particularly confusing for stationkeeping and—though unlikely—could knock a satellite out of its slot.

Fortunately, domestic and international conventions govern the placement of satellites in the orbital arc. Over the United States, satellites that use the frequency called *C-band*[2] cannot be closer than

4°; those using the higher-frequency *K-band*[3] may not be closer than 3°. Closer spacing is permitted between satellites that do not broadcast in the same band. Canada requires its C-band satellites to be 5° apart, mainly because its earth stations are much farther north. Signals to and from Canadian provinces have to pass through much more of the earth's atmosphere, so they need higher power; that forces the satellites to be spaced somewhat farther apart, to minimize interference.

Spacing requirements impose a limitation on the number of satellites that can be placed in orbit, just as zoning of a city limits the number of apartments that can be built on a site. How many open slots are there? In the best arc over the continental U.S. there is room for only 22 C-band satellites with 4° spacing, and 30 K-band satellites with 3° between them: a total of 52.

There are some ways to get around the limits. If the 4° spacing were narrowed to 2°, as many engineers have argued, almost twice as many satellites could be accommodated in the arc. But other engineers worry that closer spacing will be hard on older, existing satellites, and could render them obsolete even before the end of their useful lives. Satellites which were designed and built for 4° spacing may overpower newer ones designed for 2° spacing, and could prove to be unsuitable neighbors. In addition, closer spacing would require that the receiving equipment on earth be better able to distinguish intended signals from interfering ones, which would require their signal-collecting "dishes"[4] to be larger and more costly. It would help if they could be located directly under the satellite, that is, on the same meridian of longitude; but most earth stations are adjacent to population centers or corporate premises which cannot be moved. Closer spacing is impractical anyway for earth stations that are too far to the north or south, because higher power is needed to get signals through the air, and that might cause interference.

SPACE STATIONS AND ANTENNA FARMS

Another way out is to gang the antennas together into a giant "space station" many times larger than any one satellite could be. The idea is merely an extension of the hilltop "antenna farm." Some independent companies operate multipurpose antenna masts; their clients are local broadcasters. By engineering and mounting all the transmitters at one time they can keep them free from mutual inter-

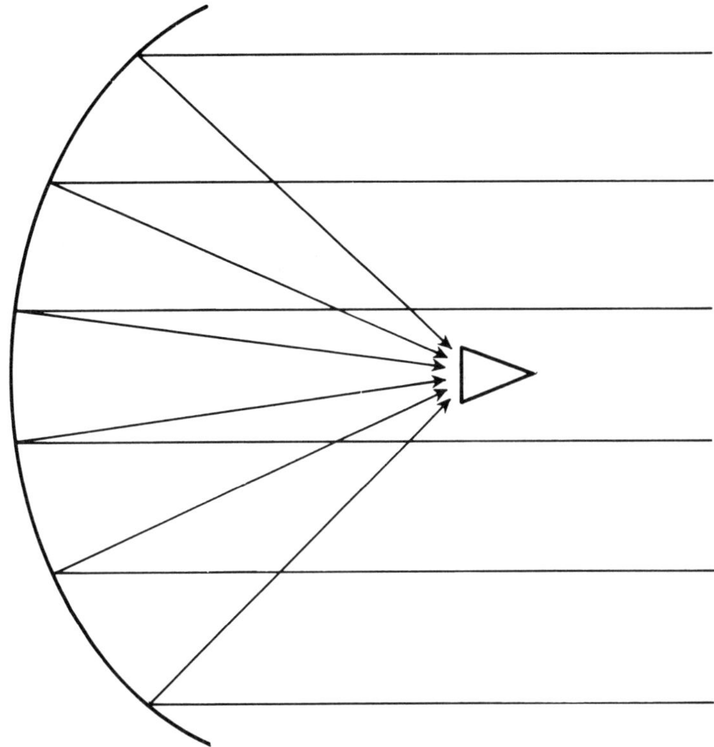

A satellite dish is (usually) a paraboloid that collects signals and bounces them into its mathematical center where a feed horn collects them and sends them to be amplified.

ference. Sharing the mast also allows the transmitters to share a common power source, which is a cost-saving advantage. The operator needs only a single permit from the local zoning authority. The broadcaster pays a share of the construction and maintenance in the form of a lease rental and trades an otherwise exclusive right to the site in exchange for simple operation and stable costs. The listening and viewing public also benefits from antenna farms: they can enjoy more kinds of programming at lower cost with a single receiving antenna pointed in a single direction, instead of having to buy several antennas and point them in different directions.

A space station would work in much the same way. Being large, it would be more stable in orbit and would use less stationkeeping energy, despite its size, than a satellite would use. A space station could be self-supporting: revenues from communication traffic

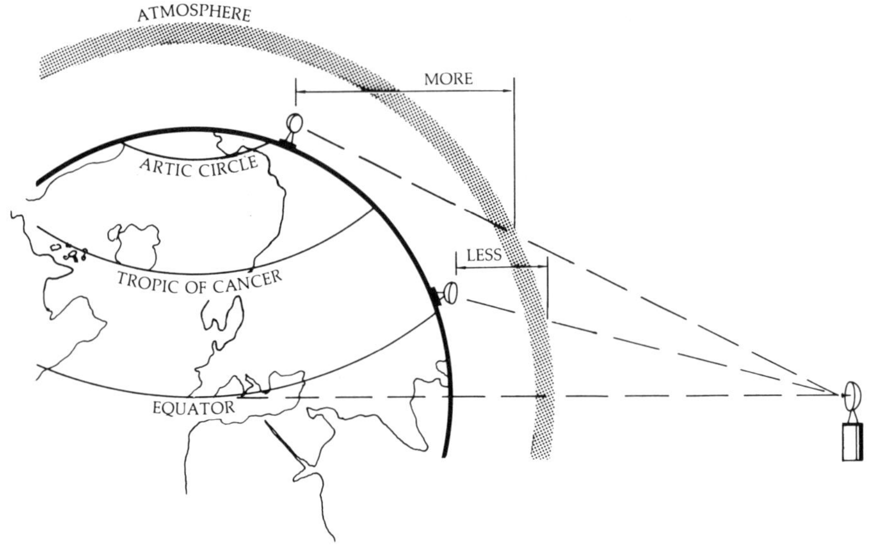

The farther from the equator an earth station is the more power it needs to send signals through the atmosphere, because there is more of the atmosphere between it and the satellite.

would pay for construction, expansion, and maintenance by (possibly) live-aboard crews. But geostationary platforms are many years in the future. They would have to be built in space, using technology that is only now being tested in experiments with the Space Shuttle. The cost—however economical in the long run—will be literally astronomical, and no one nation or corporation on earth is prepared to absorb it all. Their political or military affiliations could place them in jeopardy. Conflicts of interest between civilian and military clients, or between the public and private sectors, are far from resolved.

INEXORABLE REALITY

The fact is that orbital parking slots are limited. Each one that is filled reduces the total number. Once a car is parked in a garage, no other car can occupy that slot. Even if the lines between the cars are redrawn, the number of parked cars is ultimately limited by the size and shape of the garage itself.

A key point of this book is that, no matter how efficiently the orbital arc is filled nor how tightly spaced the satellites are, *there is only room for a finite number of satellites in the Clarke Orbit.* New technologies seem to push the limit farther away every year, but they

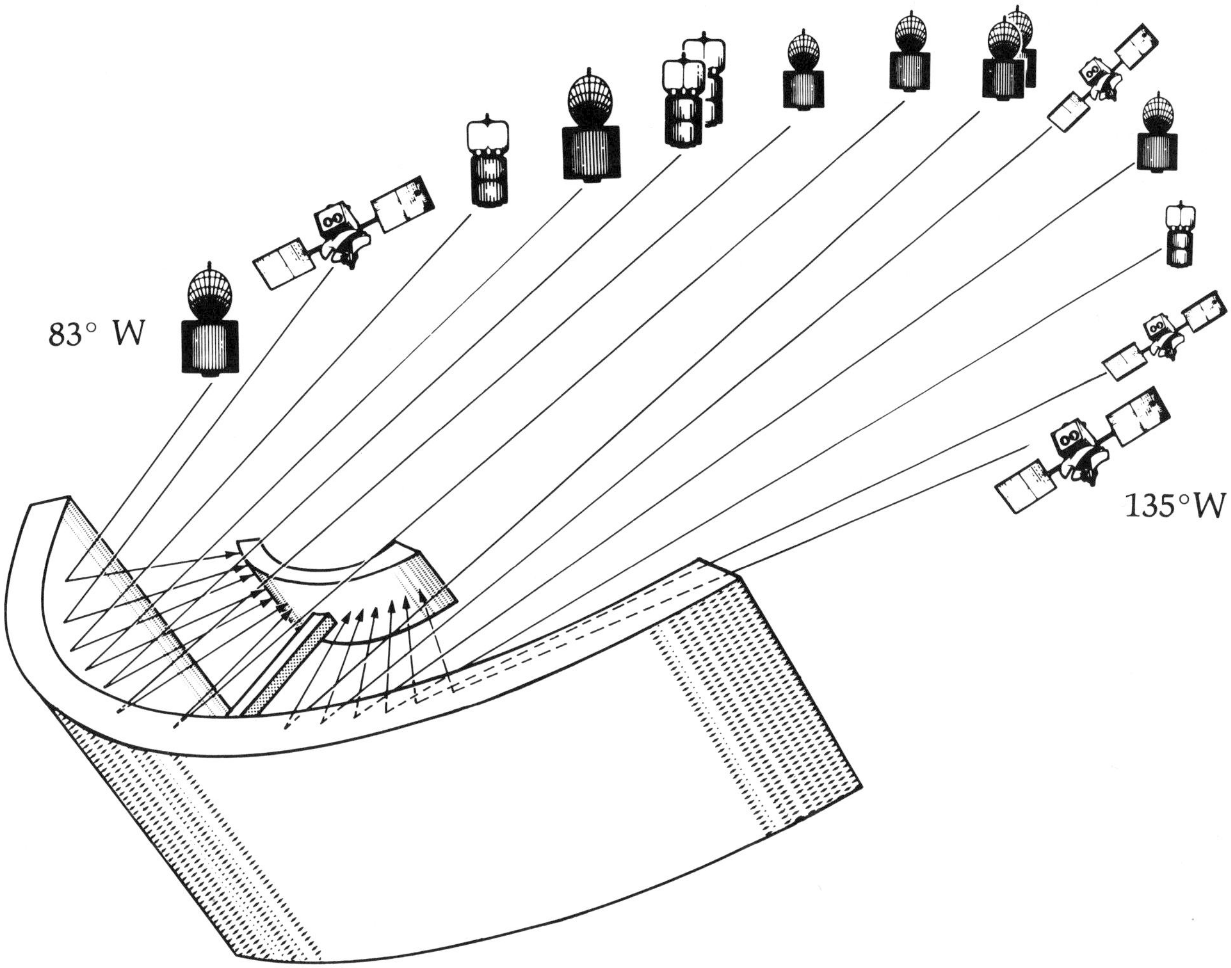

Instead of having a parabolic shape, a dish can be spherical or it can be a section of a torus (doughnut-shaped). The dish may have several feed horns in the center to accept signals from different satellites, eliminating the need to be realigned for each one.

affect the *efficiency* of the orbit's use, and the *volume* of communication which it supports. Technology alone cannot raise the capacity of the orbit to accept new satellites. In the oil business, new drilling rigs can extract more of what is underground; new refineries can distill more usable products with less waste; better-informed consumers can favor fuel-efficient machinery and appliances. But none of these improvements affects the world's inherent oil reserves. Seen in that regard, the orbital arc is a natural resource which—like oil, coal, and water—must be conserved and managed wisely.

Endnotes

1. ***Repeater antennas*** are relays. A signal from a small radio, even a hand-held walkie-talkie, may reach only a few miles, especially if it is close to the ground or surrounded by hills and tall buildings. Once the signal reaches a repeater antenna, which is usually placed high on those same hills or buildings, it can be transmitted over a much wider area.

2. ***C-band*** is the common name for a frequency range of between 4 GHz and 6 GHz, with the higher frequencies used to transmit up to a satellite, and the lower ones used for reception of the returning signals. C-band is also written as *6/4 GHz* for that reason.

3. ***K-band*** is a higher frequency range than C-band: between 12 GHz and 14 GHz, and written as *14/12 GHz.* Both C-band and K-band are used for a variety of purposes, but many commercial television programs are currently distributed over C-band, while K-band is more often used for computer data and for military traffic.

4. ***Dishes*** are satellite antennas; they are shaped like bowls to accumulate the signals that fall to earth from the satellites. The most common shape of their curve is a paraboloid (a solid figure formed by rotating a parabola on its axis), but spherical and other shapes are used for special purposes.

Chapter 3

TRADE-OFFS IN SATELLITE COMMUNICATION

Life is full of trade-offs. If your home heating bills are high, and you convert from an oil furnace to a wood stove, you'll be freed from rising oil prices, but you'll pay money for the new stove, you'll expend physical energy chopping and stacking wood, and you'll have to participate in some scheme for forest management or risk denuding the land.

In designing complex systems, such as a house, there are many alternatives from which to choose, and they tend to be mutually exclusive. The architect and the owner must indentify key decision points early in the design, and make the compromises before the house itself is actually built. It is rarely possible, for example, to move a bathroom or kitchen once a house is under construction. For absolute economy, the kitchen and bath should share a drainpipe and a waste "stack," and be fairly close to the source of hot water. But the family's style of life may dictate that the kitchen be at one end of the house and the bathroom at the other. The trade-off is one of money for convenience.

Satellites, too, have trade-offs. The decisions made on earth affect not only the technical operation of the satellite but the way it interacts with its users. The choices can *only* be made on earth, before the launch, because we do not yet have the skill to physically change, repair, or modify a satellite once it is in geostationary orbit.

A SATELLITE SYSTEM'S COMPONENTS

The receiving equipment on the ground is complex, but divided into only three basic components. There is an antenna, usually shaped like a dish or bowl, which focuses the radio waves it collects into a tiny funnel. That funnel, called a *feed horn,* pipes the signals into an amplifier. Like most electronic devices, amplifiers make "noise" of their own—radio waves that can slip in with the received signals and which will be amplified right along with them. The best amplifiers for satellite reception are called *low-noise* amplifiers: they boost an incoming signal while contributing as little noise of their own to that signal as possible. The better able they are to do that, the more costly they are to build or buy.

After the signal is amplified, it is changed into something which can be "understood" by whatever machine the signal was intended to reach. If that signal is computer data, it will be fed into a computer. If it is a telephone call, it will be connected to the telephone

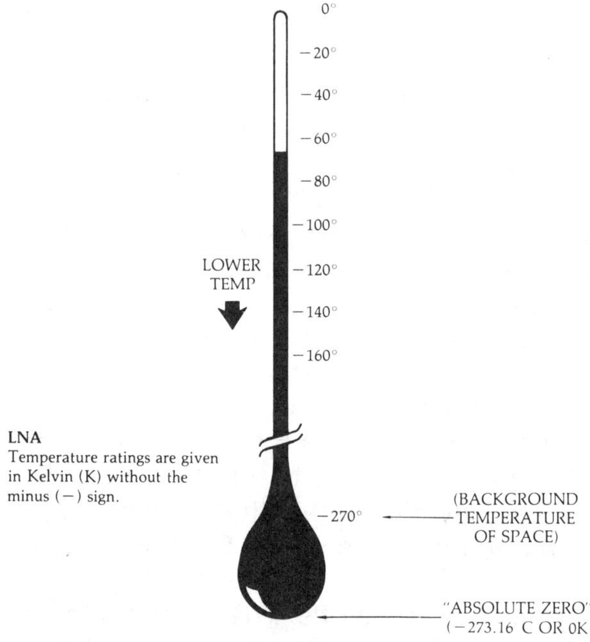

A Low-Noise Amplifier (LNA) is sensitive to the difference between the strength of a signal and background noise. It detects the added "heat" from the signal. An LNA rated at 80K (Kelvin) is not as sensitive as one rated at 90 K, and one at 120 K is even more sensitive. For any given satellite receiver, the lower the LNA temperature rating, the more the desired signal will be amplified—compared to the "noise" of the other unwanted radio signals received. "Signal-to-Noise" is a ration of amplified signal to amplified noise, and the higher the ratio the better. (e.g.,S/N of 53 is better than S/N of 49, and means that the signal is 53 times stronger than the noise.)

network. If it is a tv program, it will be converted to the kind of signal that a tv set can display; in that case, the machine that changes it is called a *converter.*

To transmit a signal, the process is reversed. A converter of some kind changes the incoming signal to what the satellite expects to receive. It is amplified by a low-noise amplifier, pushed through the feed horn, and thrown out by the dish. *Earth stations,* such as that, usually require a larger dish and more complicated equipment to transmit up to a satellite than to receive what comes down from space.

On board the satellite, there are miniature versions of the earth station equipment: a dish antenna, a feed horn, an amplifier, and some kind of device for processing the signals. Each whole system, taken together, is called a *transponder.* There are many transponders on board—24 is a common number—and each can be used both to send and to receive signals. A few transponders are spares; they are not used except as replacements for those that fail.

After more than 20 years, some aspects of satellite technology are known to be flexible, while others are not. Whenever there is a choice to be made, the engineers decide what *can* be done, and the operators decide what *will* be done. Here are some of the choices that affect both the technical and the operational sides of a satellite system. Natural limits impose most of the restrictions, but—as later chapters make clear—there are human limits, too. Awareness of those limits, and how to balance their trade-offs, is at the heart of any sensible, sensitive management program.

POWER

Communication satellites generate their own electricity on board. Light-sensitive chemicals, in packages called *photovoltaic cells,* convert sunlight into electric current; then a battery stores and regulates that power. It also supplies power whenever the satellite is "eclipsed" by the earth, that is, when it is in shadow. Photovoltaic cells are not very efficient, but in space there is no atmosphere, dust, clouds, or rain to interfere with them. They produce electricity virtually without problems and almost never wear out.

Some of the electricity is used by the satellite itself. It spins the gyroscopes that keep the satellite stable in its orbit. It operates the internal, "housekeeping" circuits. But most of the electricity goes to the antennas. The power of the satellite can be used to make the broadcast signal stronger, or it can be used to amplify the signal received from earth. A balance is usually struck which fills both needs efficiently.

A satellite is a kind of "mirror" in space, receiving, amplifying, and retransmitting whatever is sent up to it; so it is customary to speak of each pair of receiving and transmitting antennas as a single unit. The name for the pair is *transponder,* and the word refers to the two pieces of equipment as if they were one. The number of transponders of the satellites is, generally, an indication of its power.

EARTH VERSUS SPACE

Putting greater power on board the satellites lets the earth stations be small and low-powered—hence, fairly inexpensive. If the power is left on earth, in the form of large, powerful ground systems, then the satellite transponders can be fairly low-powered and less expensive.

This is a very important trade-off. Many of the conflicts between rich and poor nations over satellite resources could be resolved if

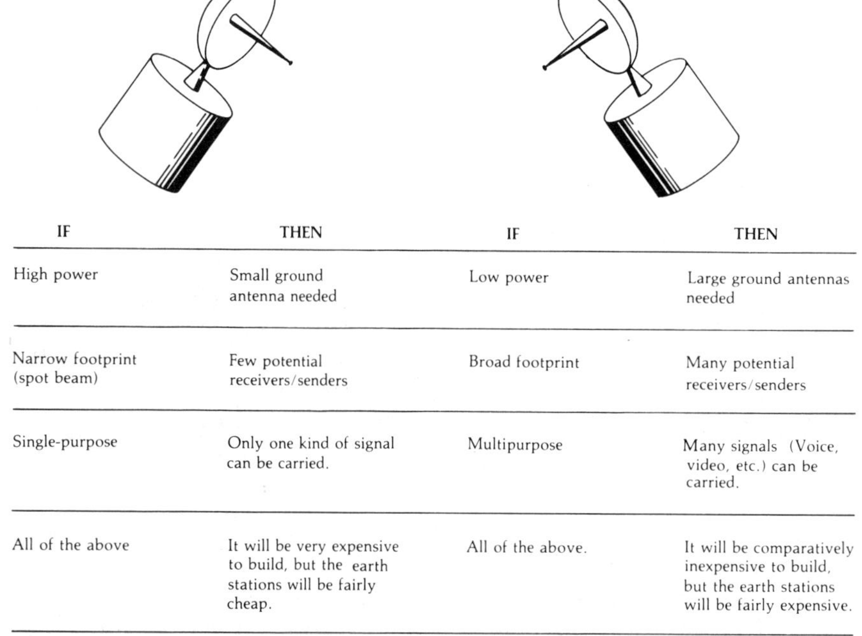

IF	THEN	IF	THEN
High power	Small ground antenna needed	Low power	Large ground antennas needed
Narrow footprint (spot beam)	Few potential receivers/senders	Broad footprint	Many potential receivers/senders
Single-purpose	Only one kind of signal can be carried.	Multipurpose	Many signals (Voice, video, etc.) can be carried.
All of the above	It will be very expensive to build, but the earth stations will be fairly cheap.	All of the above.	It will be comparatively inexpensive to build, but the earth stations will be fairly expensive.

Assume that two satellites are capable of relaying the same kind of signal.

IF	THEN	IF	THEN
The dish is large	It will pick up more signals from space, and transmit with less electrical power.	The dish is small.	It will pick up fewer signals, and require more electric power for transmitting.
The low-noise amplifier is powerful (Rated lower temp. K).	The dish can be smaller and cheaper.	The low-noise amplifier is not so powerful (Rated higher temp. K).	The dish must be bigger and more costly.
The dish is close to the center of the footprint.	The dish can be small and the L.N.A. need not be powerful.	The dish is far from the center of the footprint.	The dish must be larger and the L.N.A. more powerful.
All of the above	Transmission and reception will be easy, reliable and of high quality.	All of the above	Transmission and reception will be difficult, unreliable and of low quality.

Assume that two dish antennas are capable of relaying the same kind of signal.

there were agreement on the power installed in the satellites. NASA's first communications satellite (ATS-1) was so powerful that users needed only a hand-made antenna that was not much larger than a home tv antenna to gain access to it. By putting the power aloft, the cost and power of the ground equipment were kept to a bare minimum (see Chapter 8). Similarly, ATS-6 was NASA's first attempt at broadcasting tv from space (see Chapter 17). When it was placed over India, rural villagers made antennas out of chicken wire that were adequate for reception, because the power was aboard the satellite.

By contrast, the antennas on INTELSAT's first commercial satellites were not nearly so powerful. It took an earth station dish 30 meters in diameter to catch the signal, and it took a very expensive and sensitive amplifier to make those signals intelligible. The 30-meter installations are still in use, and each one costs millions of dollars. In fairness, however, the early NASA satellites had their own limits: ATS-1 could not handle more than a few telephone calls at once, and ATS-6 could broadcast only one tv program at a time, to just a few selected places on earth. INTELSAT's first satellites, by comparison, carried hundreds of telephone calls *plus* tv transmissions, which could be picked up virtually anywhere on earth that had a line of sight to the satellite. The higher costs were shared among many more users.

SIZE

Generally, the larger a satellite is, the more it can do. Bigger solar panels give it more on-board electricity, so it can carry more transponders, which means it can serve more customers. Larger fuel tanks for the stationkeeping rockets give it a longer life in orbit. Its many delicate systems are duplicated so that, if one fails, a replacement can take over immediately; this redundancy, however, also adds weight.

But a large satellite is more expensive than a small one, and higher construction costs require higher insurance premiums (see Chapter 11). The rocket that launches it must be correspondingly larger and more powerful than one for a smaller satellite. That, too, costs more money, and the launch may have to be delayed until a suitable "launch vehicle" is available.

The maximum size of the satellite is usually determined by the carrying capacity of the rocket: the price of the launch is more or less the same no matter what the payload, up to that maximum capacity. But in the 1980s, the American Space Shuttle may change the economics of launches. On its fourth flight it carried two satellites to low-earth orbit, one for Satellite Business Systems and one for the Canadian government. With laudable precision, the two birds were slipped out of the Shuttle's cargo bay. Each of them carried a small, self-guiding, computerized rocket called an *inertial upper stage* (IUS) which boosted them up to geostationary altitude and helped orient them in position. Because the Shuttle is subsidized by NASA and the American public, the actual cost to SBS and Canada was lower than a comparable launch by conventional rocket. No one is predicting when, but inevitably NASA will begin to charge cost-plus rates for Shuttle launches as part of its "commercialization" project (see Chapter 15 on Landsat).

Other countries subsidize launches too. Banks in France and others affiliated with the European consortium called Arianespace offer modest rates of interest on loans to pay for satellite launches by Ariane vehicles. And soon there may be private-sector competitors. One is Space Services, Inc. of America (SSI), which test fired its first rocket in 1982 from an island off the coast of Texas. Dubbed "Conestoga" after the covered wagons that helped colonize the American west, the SSI launch vehicle is theoretically capable of taking something into low-earth orbit. SSI hopes to get some "pay" out of its "payload" by the middle of the decade.

SHAPE

The shape of a satellite is determined early in its life, in response to the need to keep it pointed at roughly the same place on earth. Deviations from its pre-set orientation are sensed by a computer, and can be adjusted by telemetry. A satellite has gyroscopes inside it and, as everyone who has played with gyroscopes knows, whatever is holding the gyroscope tends to spin in the opposite direction. The inevitable equal and opposite reactions to the gyros must be damped.

So one of two engineering feats must be undertaken. Either the satellite must be made to spin *under control* in the opposite direction from the gyroscope, or the gyro must be isolated, mechanically, so that it will not cause motion in the satellite. There are two basic configurations. The barrel-shaped satellite spins counter to the gyroscope, and is called a *spin-stabilized* satellite. The one with wings has the gyros spinning inside a sealed container, and has what is known as a *body-stabilized* design. Hughes Aircraft favors the spin-stabilized barrels; RCA tends to build winged, body-stabilized satellites. Neither design is significantly "better" than the other, and the trade-offs between them are fairly technical.

The outer skin of the barrel-shaped, spin-stabilized satellites is covered with photovoltaic cells; rotation exposes all of the cells to sunlight. The "paddles" on body-stabilized satellites are its solar panels; mechanical adjustments keep them pointed toward the sun. There is no apparent difference between the two designs as far as communications is concerned. INTELSAT, for example, has purchased both kinds of satellites and has not experienced any difficulty integrating them into its system.

In any case, each design can be customized to meet different needs. Hughes Aircraft's HB-376, for example, has been the "best-seller" in its field: AT&T: Satellite Business Systems; the Canadian, Australian, and Indonesian governments; Western Union and Hughes' sibling, Hughes Communications, have all bought them. The customizing process was recently illustrated when AT&T bought several HB-376s and called them Telstars. For example, of the "standard" 24 transponders aboard, AT&T chose to reserve six as spares, so that if some transponders fail, there will be enough left to take their place. AT&T predicts that, after seven years, 22 transponders will remain active. By contrast, the satellite which it replaces—called Comstar—activated all 24 transponders after launch, and so had a greater circuit capacity right away, but after seven years, only 18 transponders were still working.

Trade-offs with on-board power were part of that design process too. The on-board power of Telstar, 670 watts, is not much greater than that of Comstar (600 watts), but it is divided into fewer transponders (18, compared with 24) at the start of its life, and is thus somewhat more powerful and efficient.

FREQUENCY

Hertz is the number of "waves" per second that radio propogation generates.[1] The frequency at which a satellite receives signals is called the *uplink,* and the frequency—never the same—at which it transmits back to earth is the *downlink.*[2]

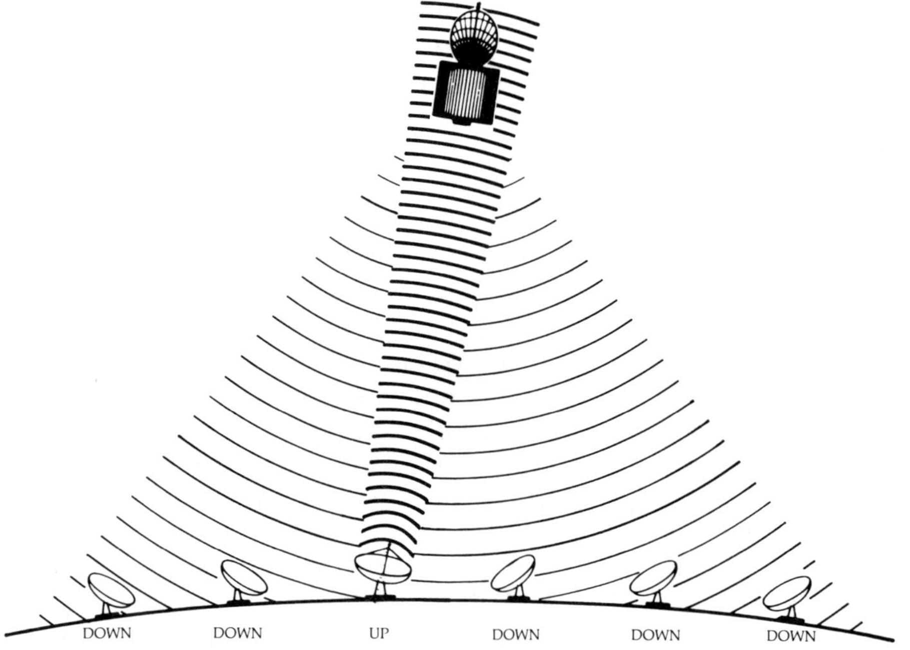

Uplink and Downlink—antennas that transmit up to the satellite are called uplinks. Antennas that only receive signals and do not transmit them are called downlinks. Those that only receive television programs are called TVRO dishes (Television Receive - Only). And are the dishes that people install in their backyards.

The uplink frequency is always higher than the downlink frequency and is usually listed first when the two are placed together; e.g., "6/4 GHz" means that uplinks are approximately 6 billion Hz (6 gigahertz) and downlinks are approximately 4 gigahertz.

To get an idea of the scale of frequencies, consider High Frequency (HF) radio, which is popularly known as "shortwave" radio because the higher the frequency, the shorter the "wave length." An electrically charged layer of the atmosphere, called the *ionosphere,* acts like a mirror for radio frequencies that are at or below the HF band. Any higher frequency tends to go right through the ionosphere and be lost in space.

The frequencies on ATS-1 are in the Very High Frequency (VHF) radio band, which is the band used by city police or fire department radios. The bandwidth in that VHF frequency range is capable of carrying two-way simultaneous voice communication, but so that more people can use it at the same time, it is almost always used in one direction at a time; that is why one person will say "Over" to the other when he or she is ready to listen. Since the 1930s, VHF has been used for television broadcasting; viewers know it as the band with channels numbered 2–13.

ATS-6 used the Ultra High Frequency (UHF) band, which is now familiar in the U.S. as "the other dial" on tv sets, selecting Channels

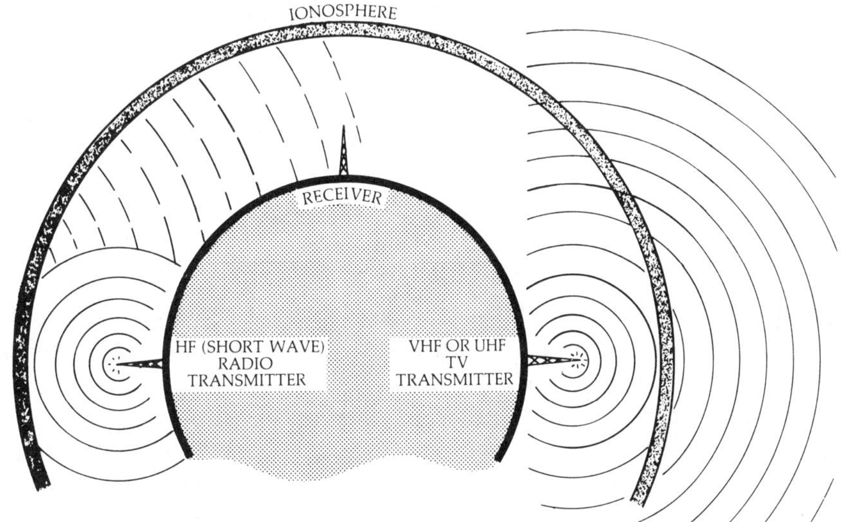

Either VHF or UHF transmissions are limited to line-of-sight communication within the horizon. Radio signals in the HF and LF (low frequency) range can bounce off of the ionosphere and be received far away around the world.

14–83. The UHF, being a higher frequency range than VHF, has different characteristics from VHF. For example, VHF signals are less susceptible to interference from hills, buildings, and other obstacles than UHF signals are, but the range of UHF over relatively flat terrain is greater than that of VHF. And UHF needs more electrical power, but VHF is more wasteful of spectrum—UHF broadcasting leaves more "room" in the band for other users than VHF does, and is therefore a better conserver of the spectrum.

Why use higher frequencies? Because they squeeze through much more information per second than lower frequencies do. Think of Morse Code for a moment. Suppose a telegrapher can send 100 words per minute. Doubling or tripling that speed would produce a two- or three-fold increase in "throughput," which is the volume of information carried through a circuit. If, *hypothetically,* each "cycle" (hertz) of eletricity could carry a single letter, then a 60 Hz household current could carry 60 characters per second; a 144 MHz police radio could carry 144,000,000 and a 6 GHz satellite transponder could carry six billion characters every second. The numbers are exaggerated but the *comparison* is not. The trade-off for efficiency is this: higher frequencies of transmission can carry more information than lower frequencies do, over any given unit of time.

Transponders using a 14/12 GHz band can carry more information per second than those using a 6/4 GHz band. But at any one level of transmission power, there is a trade-off between the frequency and the breadth of its range. If a higher frequency is used, it must be concentrated into a narrower beam for reception, or else it must use more power than a lower-frequency signal to reach the same wide area.

BANDWIDTH

Although domestic satellites are used mainly to carry television programs, most international satellite traffic comes from telephone calls. Despite the appeal of televised boxing and soccer matches broadcast around the world, ordinary voice telephone communication is the largest market now served. But, since the 1960s, it has been considered impractical to devote a whole transponder to carrying only one voice in only one direction. Today's INTELSAT transponders are "wide" enough to accommodate thousands of voice circuits *in both directions at once.*

The use of the word "wide" is intentional—it conveys the idea of a highway with many lanes, which is a good analogy for "bandwidth." First, think of a pair of thin wires, such as those running to a home telephone. That "wire pair," as it is called, can carry one voice circuit. Ganging a few of them together into a cable would let them carry more conversations, but they would create electromagnetic interference among themselves. Coaxial cable—used by cable television systems to run tv into people's homes—is a more scientifically efficient way to gang wires together. Those tv cables *could* carry a few hundred telephone calls, but cable tv operators have not pressed hard to explore that opportunity. Local regulations, opposition from the entrenched telephone company, and the lack of enough customers for an economy of scale can all discourage cable tv operators from offering such a service. Still, the bandwidth of coaxial cable is great enough for carrying as many as several dozen television channels at once, and each tv channel is "wide" enough to carry a few thousand telephone calls.

The point of discussing bandwidth is that the satellite operator must decide early in the bird's design just how many transponders it will hold. The designer will try to mount as many transponders as possible, but there are limitations based on a satellite's on-board power, and practical considerations with respect to the users.

Hertz is also used to describe the exact "width" of a communication channel. The bandwidth of a satellite transponder has only recently been made wider than one tv channel (36 MHz). It would be ideal, for the operators, if each transponder served one earth station. But users (their customers) prefer flexibility, and flexibility eats into the bandwidth. Remember that a highway can fork apart into various smaller roads at its destination, but each can handle only a fraction of the primary highway's traffic, and each needs curbs, gutters, and median strips to keep cars from sliding into one another.

FOOTPRINTS

The common term for the area of coverage under a satellite is its *footprint.* In theory, a footprint would resemble the pattern a flashlight beam makes: circular, with the brightest spot in the center, and the light tapering off toward the edges. But because the earth's terrain is uneven, in motion, and subject to changing atmospheric conditions—and because the satellite is not directly "over" the northern or southern hemisphere where it is most likely to be used—the actual footprint of a satellite looks more like a . . . footprint. Signals from a typical satellite that cover the continental United States, for example, will not be evenly distributed; the operators may direct it to concentrate in the eastern states, or in the midwest, depending on where the customers are.

It is considered impractical to build a satellite whose coverage of so large an area would be entirely uniform, because the strength of the signal falls off toward the edges just as light from a flashlight beam does. Instead, users acquire larger dishes to accumulate more of the signal, the farther from the center they are located.

The footprint of a high-frequency transponder can be made smaller than the footprint of a lower-frequency transponder, operating at the same output power. C-band transponders are widely used for distributing television programs to tv stations around the U.S. because virtually everywhere in North America is in the enormous footprint of the C-band. The SBS satellites, using the higher frequency K-band for business communication, have a wider bandwidth in each

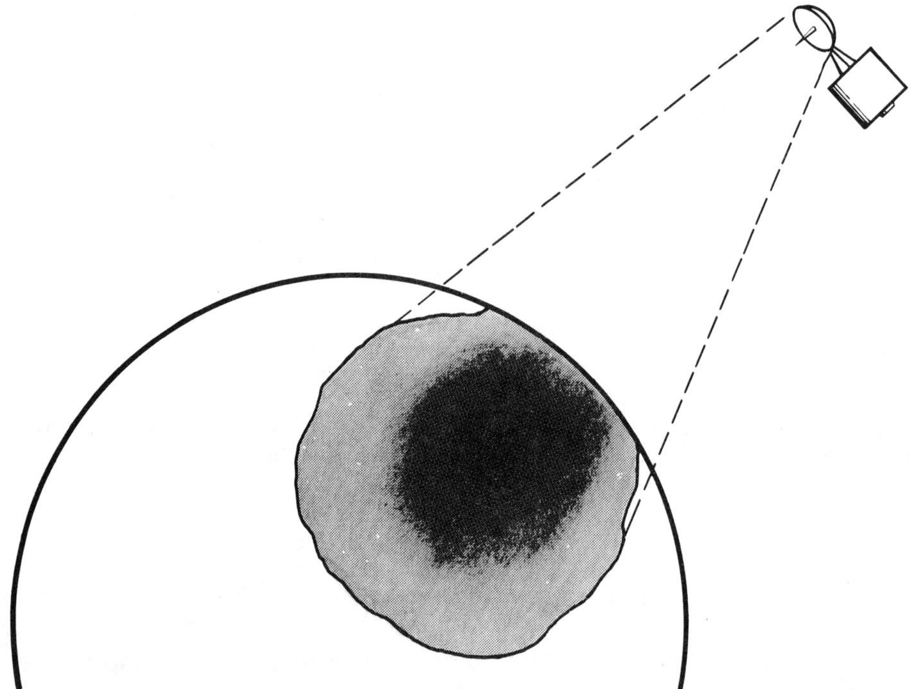

The theoretical footprint of a satellite is like a flashlight beam—circular and concentrated at the center. The intensity tapers off toward the edges. However, the curvature of the earth, the thickness of the atmosphere, and the design of the satellite antenna make the *actual* footprint have an irregular shape.

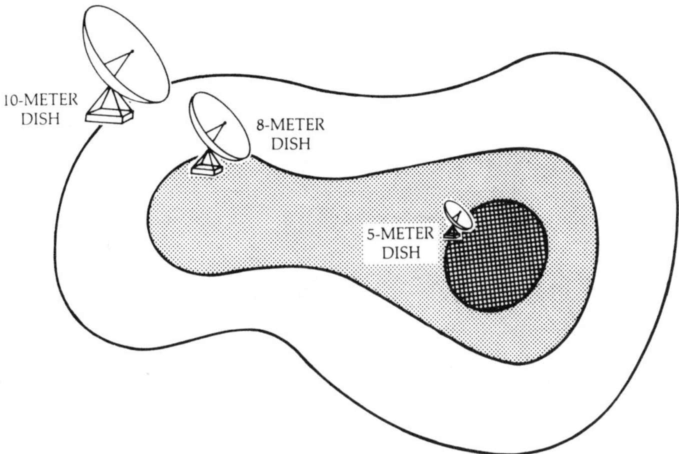

An earth station farthest from the center of the footprint must have a larger dish than one that is nearer because the intensity of the signals at the edge is weaker than that near the center.

transponder than the C-band satellites do; footprints of their transponders cover only about the area of the continental U.S.

When direct-to-home broadcast satellite (DBS) service transmits tv to the U.S., it will be using the Ku-band (pronounced "k-u"), which is of a still-higher frequency. Although they could build transponders with nationwide footprints, they will not: instead, they will use four separate satellites, one to cover each U.S. time zone. There are two reasons for that. First, local broadcasts can be made at the appropriate local time, and second, there is a trade-off between power and the size of the footprint. A satellite transponder that could send and receive Ku-band signals anywhere in the continental U.S., would require enormous power—possibly even greater than the photovoltaic cells of the satellite could generate. The trade-off that DBS providers will make is to keep the size of the footprint small, and thus conserve electricity.

Transponder antennas with very small footprints are called *spot beams;* they are used to augment coverage. There are spot beams, for example, to reach Hawaii and Puerto Rico, so no power is wasted covering the oceans that separate them from the continental U.S. Some satellites use motors to "steer" transponder antennas toward specific areas; *steerable beam* antennas can be moved to follow changes in demand. Some experiments have also been made with

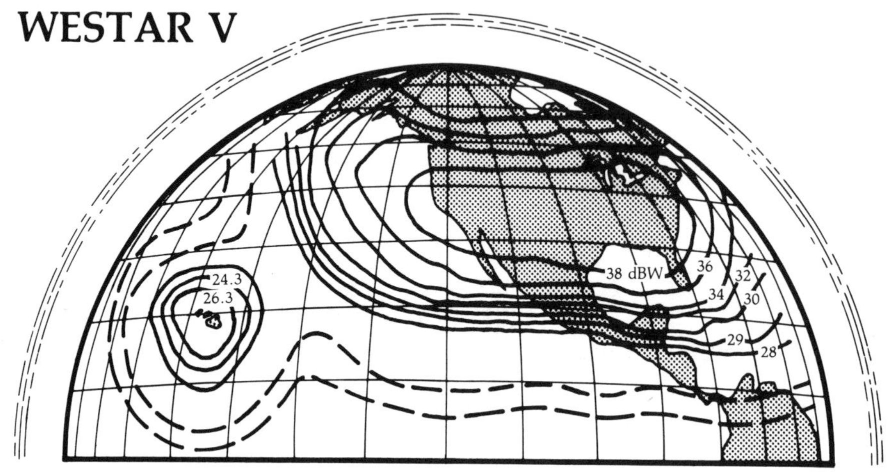

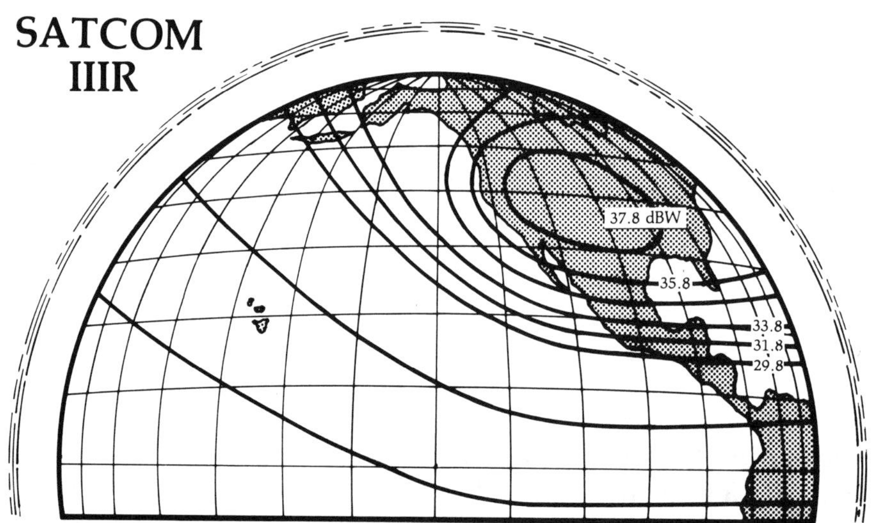

Footprints from Westar V and Satcom IIIR show how the signal strength gets weaker outside the main focus of the satellite antenna. The numbers are technical parameters; the higher the number, the stronger the signal. Within the continental U.S., signals from both satellites are approximately the same strength, but because Westar V is positioned farther east (at 123° W) than Satcom IIIR (131° W), the eastern states enjoy slightly stronger signals—and therefore can use smaller or less-expensive receiving antennas and equipment. Westar V has a spot beam to serve Hawaii, although its signal strength there is nowhere near as great as that which serves the mainland, so Hawaii residents need a larger and more expensive antenna system than do their mainland cousins.

small-footprint antennas that scan over a large area, "interrogating" different ground stations to see if they have any uplink signals to send, and then focusing their spot beams onto those areas.

MAN-MADE INTERFERENCE

The first commercial communication satellites used frequencies of about 1.2 GHz, which—unfortunately—was very close to the frequencies that terrestrial microwave systems use. That meant that a receiving dish was just about as likely to pick up local traffic as it was to get what the satellite was transmitting; even worse, the local traffic came from a source much closer to the receiver, so it tended to overwhelm the signal falling from the sky. Very quickly, new satellites were built to use C-band (6/4 GHz) and even higher 6 and 8 GHz.

Near cities, however, interference from local microwave networks—some of which operate on frequencies that are near the C-Band—can sometimes be too great for reliable satellite communication. Putting earth station dishes away from the cities, however, means that the majority of customers cannot have one of their own. What's more, the operator has to string a cable, or align a microwave link, from the earth station to the city to serve those urban customers. The local regulatory agency may discourage such a link because it could compete with the local telephone company, but independent links are more likely to be simply uneconomical, compared with a cooperative agreement in which the satellite operator pays a fee to the phone company for the use of a trunk line to the city.

THE "TELEPORT" SOLUTION

The Port Authority of New York and New Jersey is building what it calls a "Teleport" on Staten Island (away from most of the city's microwave traffic), with a cable under the harbor to reach the World Trade Center building in Manhattan and other business districts in the area. The city is completely socked in by a fog of radio signals and radio noise, so putting a satellite antenna in the middle of it is virtually impossible. Another "engineer's nightmare" is under New York's streets. There is a tangle of telegraph and telephone cables, some of them 50 or more years old, which give off tiny electrical currents that can also cause interference to new and well-shielded cables. The Teleport will send its signals through a glass "fiber-optic" cable that uses light instead of electricity; it is very efficient, and also oblivious to electromagnetic interference [see Chapter 16].

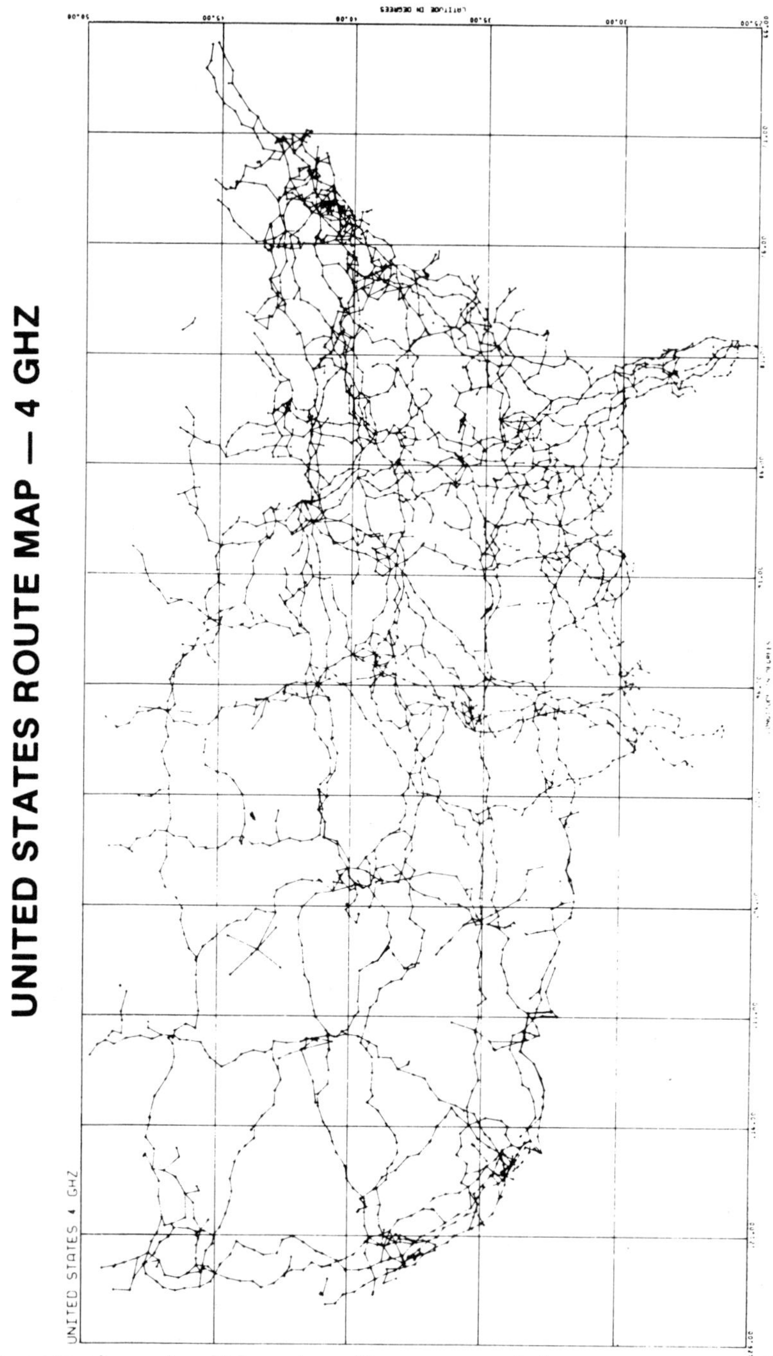

Microwave channels within metropolitan areas and along routes connecting them have become so congested that satellite transmissions in those frequencies (6/4

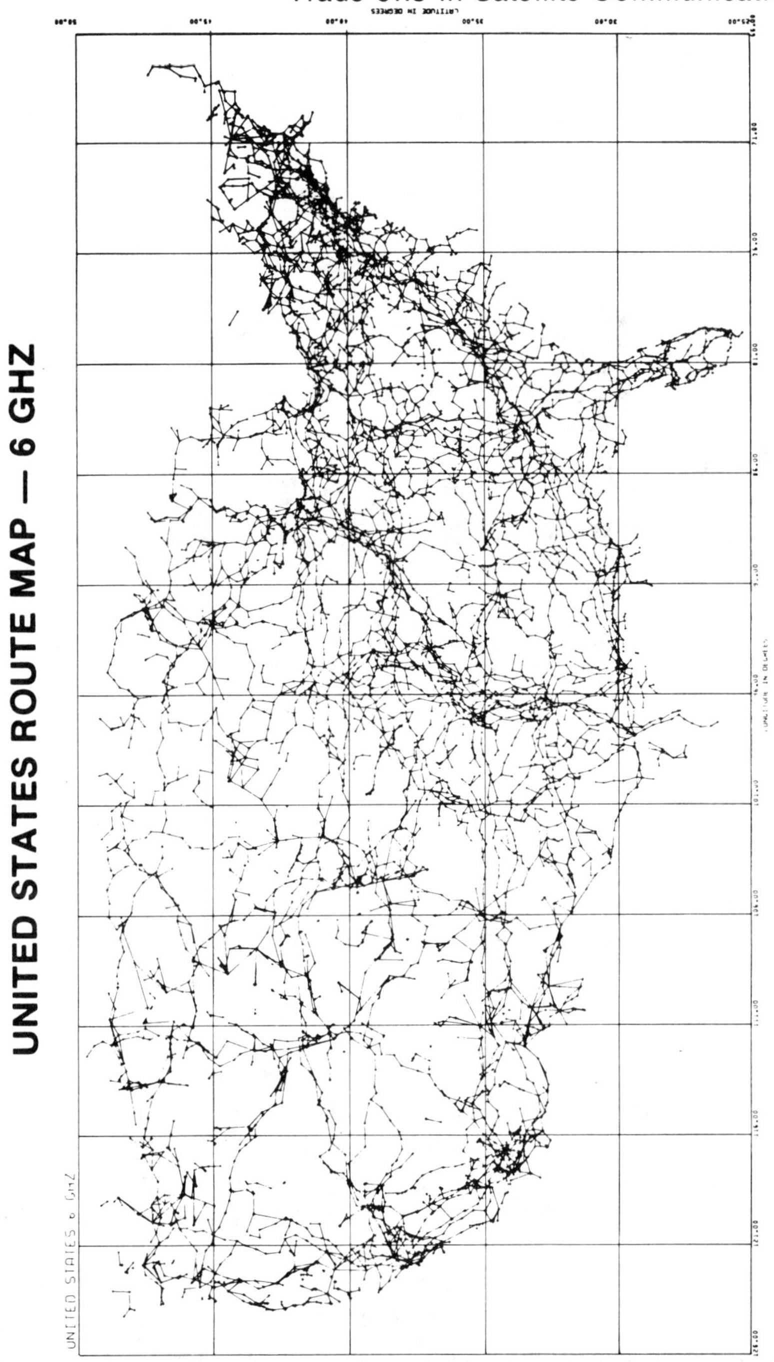

GHz) are subject to interference. This is the reason that many new satellite services are assigned to higher frequencies. (Courtesy Comsearch, Inc.)

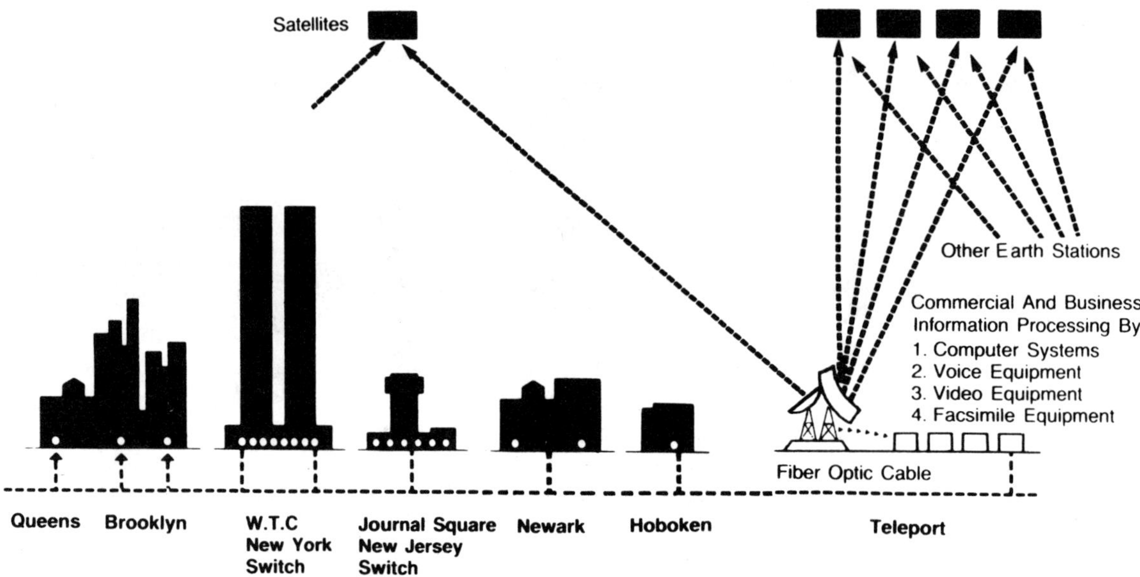

The Teleport, a satellite communication center using fiber optic data links, will begin offering service in the New York metropolitan area early next year.

Source: The Port Authority of New York and New Jersey.

A schematic representation of the "Teleport" serving the New York/New Jersey metropolitan area. A fiber-optic cable under New York Bay overcomes local interference from microwave circuits.

Financing for the Teleport comes from Merrill Lynch, the brokerage house, and technical construction from Western Union. When it is finished, some 17 earth station dishes there will have access to 22 domestic satellites and there exists the potential for direct access to international satellites, though that is not presently permitted by the FCC.

The Teleport, according to Port Authority chairman Alan Sagner, "will be especially attractive to major banks, brokerage houses, insurance companies, news organizations, and broadcasters." These customers, then, will have to weigh the trade-off between freedom from local interference and having total control over one's own facility. They will also have to balance the lease or rental of the Teleport service against the cost of building and staffing an earth station of their own.

COMPUTERS: THE DIGITAL ALTERNATIVE

For distribution by satellite, television programs do not have to be processed in any difficult way: they are transmitted to the satellite in much the same way that they would be broadcast (at a much lower frequency) through a terrestrial antenna. The downlink signal falls more or less equally throughout the footprint, so some people who live far from the cities have installed dishes and amplifiers so they can receive those satellite television programs directly. [See Chapter 5].

Another principal use of satellites is to send voices, and the future of satellite communication lies in their ability to send computer data. Voices, like television programs, are "analog" signals: changes in the transmission wave are exactly proportional to the changes in the sound of the voice, or the background and the image or color of the picture. Computer data, however, is "digital." That means that there is no direct correlation between what is transmitted and what is supposed to be reconstructed from that transmission when it is received. It is in a code which the receiver must "understand." Digital signals are simply pulses, like flashes of light or bursts of electrical current. Over a unit of time, the *presence* of a pulse can mean "one" and the *absence* of a pulse can mean "zero."

A detailed explanation is outside the scope of this book[3], but there is a trade-off between analog and digital signaling: *digital signals are much more efficient carriers of information than analog signals are, but they require greater bandwidth to accomplish it.* That means that satellites used for digital communication must use the highest frequency bands to be efficient.

K-band and Ku-band transponders are much more useful for transmitting digital signals than C-band transponders are because their bandwidths are wider. Computer data can thus be sent very easily. Voices, too, can be fed into a computer and converted into digital signals, which can then be carried by a digital satellite system. Also, tv pictures, facsimile images, and other forms of communication can be cost-effectively "broken down" into digital signals and mixed into a stream of data. That stream, then, is disassembled at the receiver and converted back into voice, tv, facsimile, etc., according to embedded codes. The high frequency permits the earth stations to be rather small. SBS, the Satellite Business System, uses bandwidths of 54 MHz (for what is called "broadband data services"), compared with the 36-MHz bandwidths of C-band analog television transmissions.

Another important reason why digital transmission is preferred over analog is that digital signals can be checked for errors and corrected much more easily. All-digital systems use sophisticated electronics at both ends to do the error-checking and correcting, in addition to the circuitry for analog-to-digital conversion, signal-mixing, the separating-out of different signals and their conversion back to analog form. So, while a higher frequency system is more efficient for digital communication than a lower frequency system is, and allows for smaller earth stations, it requires more expensive receiving and transmitting equipment—literally, a computer at each end.

NATURAL INTERFERENCE

By using Ku-band (14/12 GHz) for a single television channel, experiments have shown that the size of the earth station can be very small indeed. Direct satellite-to-home broadcasting may not be acceptable to the general public unless and until the customer's dish is only about one meter wide, and can be installed on a rooftop with no greater difficulty than installing a conventional tv antenna. But in that frequency band there is a catch: the weather becomes a limiting factor.

SBS's biggest worry was that rain would interfere with their transmissions. In rain or fog, daylight dims and everything gets dark because light waves are scattered by raindrops; fewer of them reach the eye. Extremely high frequency radio waves (especially as they approach the incredibly high frequencies of visible light) are more apt to collide with molecules of water in the air than are lower-frequency radio waves. With analog signals, such as voice or tv, there may be a momentary silence or blackout. But for digital transmissions it's worse: any interference can upset the timing—and hence the integrity—or the data.

SBS got around this problem by "redundancy." During bad weather, their computers automatically retransmit data any time that a receiving computer tells a transmitting computer that a transmission was interrupted. Error-detecting codes are built into the data stream that perform mathematical operations on the data; if the numbers aren't the same at both ends, the receiver signals the transmitter to send it again—that's how the computers "know" that a mistake has been made. Of course, redundancy is expensive, but SBS's engineers assert that, in practice, interference from bad weather is not as bad as it was predicted by theories.

Attenuation is the technical term for stopping or interfering with light or radio signals, and the phenomenon is widely used as a scientific tool. Chemists can analyze an unknown substance by burning it in a machine called a *spectroscope.* The light from combustion is passed through a prism, but certain colors are blocked because those frequencies are absorbed by the test substance; they show up as black lines in the rainbow, and their position in the spectrum tells the chemist what elements or compounds they are. Similarly, astronomers can send a radio signal to a planet and, by noting which frequencies are absorbed and which are reflected, can determine what gases are in its atmosphere.

Water vapor, oxygen, and carbon dioxide attenuate radio waves at certain frequencies, so it is unlikely that satellite networks will be able to use those frequencies, however desirable they might be otherwise. Laser light, for example, is very efficient for sending information because its frequency is so high, but beyond a few hundred feet, swirling city dust or "heat waves" rising from hot pavement distort the atmosphere and make laser transmissions through it practically useless. If the laser light is channeled through glass fiber cables, however, it fulfills its potential beautifully. Of course, the trade-off is that it can only be used where the cables are strung.

SOMETHING FOR NOTHING

Don't take that expression—something for nothing—literally: there is *always* a trade-off. But some electronic problems have been overcome seemingly without penalty. The limits to telecommunication have always been pushed outward by the need to squeeze more than one signal into a single channel. It was first done in the nineteenth century: Alexander Graham Bell himself invented a *multiplexer* for telegraphy which could send several messages over a single wire. Today, we take it for granted that the telephone wires outside our homes are carrying all our neighbors' conversations at the same time as ours, without interference.

Divide and Conquer

In propagating radio signals through the air, problems of interference are acute; the history of broadcasting is rife with complaints about "jamming," both accidental and intentional. With the development of satellites, the economics of scale demanded that a way be found to minimize interference while filling up the transponder circuits with the largest possible number of users.

Two different methods evolved in telephony, both of which are common in satellite communication today. Different circuits can be divided by *frequency* or by *time.*

"Frequency-Division Multiplexing" (FDM) means that each telephone conversation is carried on a slightly different frequency. This is fairly easy to set up, and easy to understand—most cable television services come into the home this way: many different channels can be multiplexed into a coaxial cable because the bandwidth of that cable is "wide" enough to accommodate them. The limitation on FDM is that a chain is only as strong as its weakest link—that is, while the satellite transponder and the giant earth stations can keep those frequencies separated, narrower "pipes" such as older telephone networks and household "wire pairs" cannot. But, being an analog service, FDM fits comfortably into existing analog telephone-switching networks, and is therefore very convenient to set up. That is partly offset by the high cost of sensitive *filters* which are required to keep those frequencies separated.

"Time-Division Multiple Access" (TDMA) is a way of sending many different signals over a single frequency by sending just a fragment of each, serially, in a tight stream. It's very complicated, but here is a simple analogy. If you were to print three messages on strips of ticker-tape, and cut them into pieces so that each piece contained one word, you could assemble from those pieces a *new* tape in which every third word came from a single message. If you then wrote a code at the beginning of the tape that told the reader how many messages there were, and which one came first, he or she could cut the new tape apart and reassemble the three originals.

Many messages can thus share the same frequency, and the time required to send each one is divided equally among them. Naturally, this is much more complicated than assigning each message a unique frequency because, in addition to the messages themselves, "housekeeping" or "decoding" information must also be sent along the channel. Until the cost of computers declined in the last five

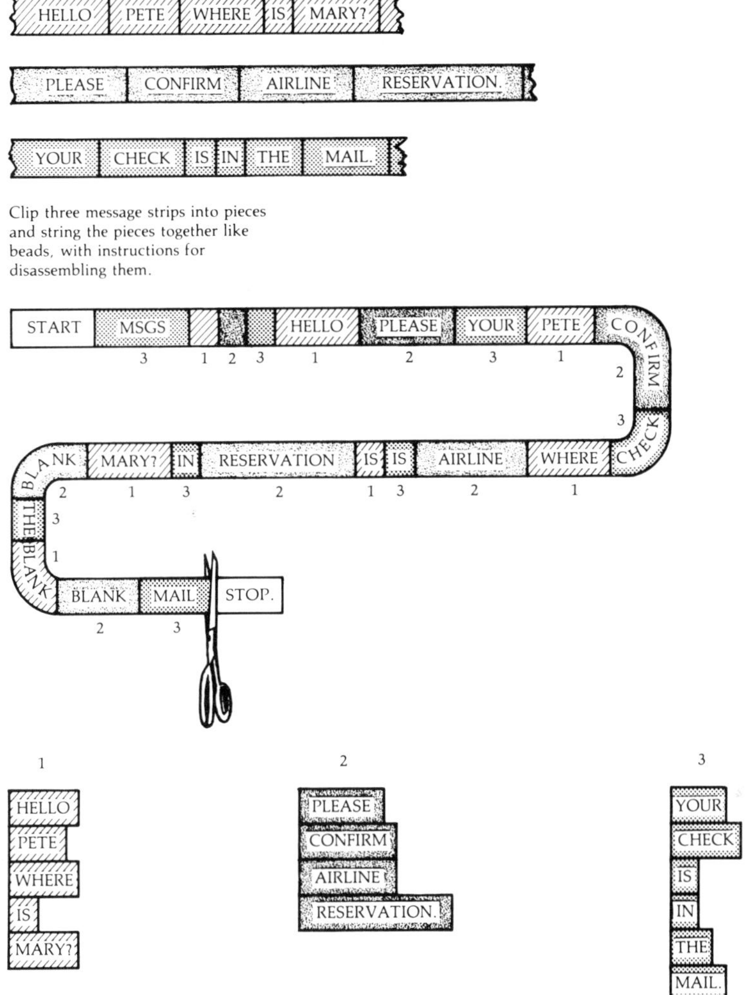

TDMA (Time-Division Multiple Access)–How it works. Example No. 1: Clip three message strips into pieces and string the pieces end-to-end like beads with instructions for disassembling them.

years, TDMA was much more expensive than FDM. Now, however, the hardware (computers) and software (the programs that tell the computers how to cut and paste the messages) are widely available and—though costly—are much cheaper than they used to be. Virtually all new satellites use some form of TDMA.

For analog signals FDM is more efficient, because the signals require no advance processing or subsequent decoding, only a sim-

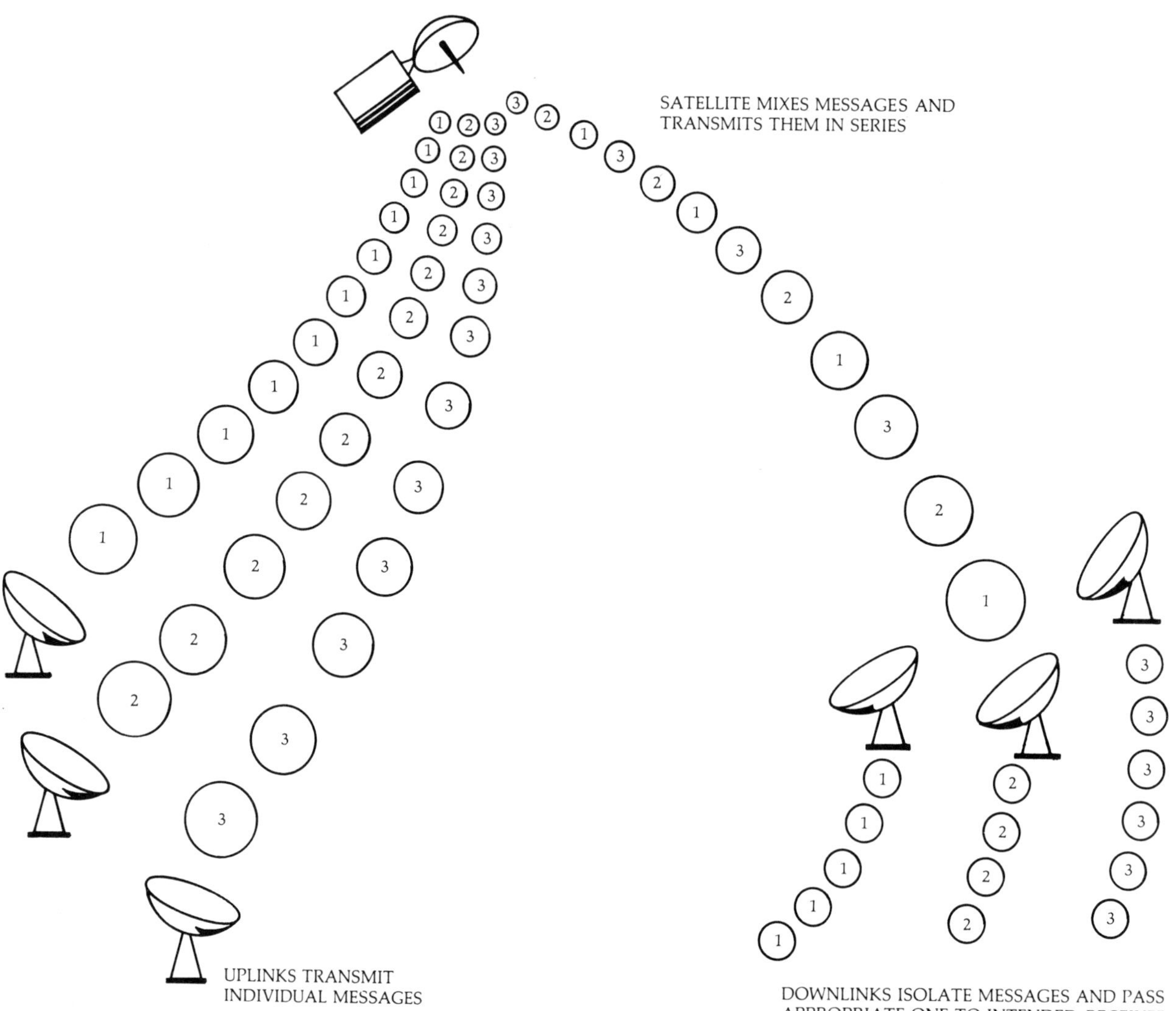

TDMA–How it works. Example No. 2: The uplinks transmit individual messages to the satellite which mixes the messages and transmits them in a series. Downlinks receive and isolate messages and pass the appropriate one to the intended receiver.

ple conversion to the FDM transmitting frequency and a conversion back when they are received. The technique, though now refined, has been common practically as long as radio has been around. TDMA, however, is better for digital signals since its short bursts are—in reality—computer data. TDMA is more sophisticated than FDM, but as more and more communication comes to be made digi-

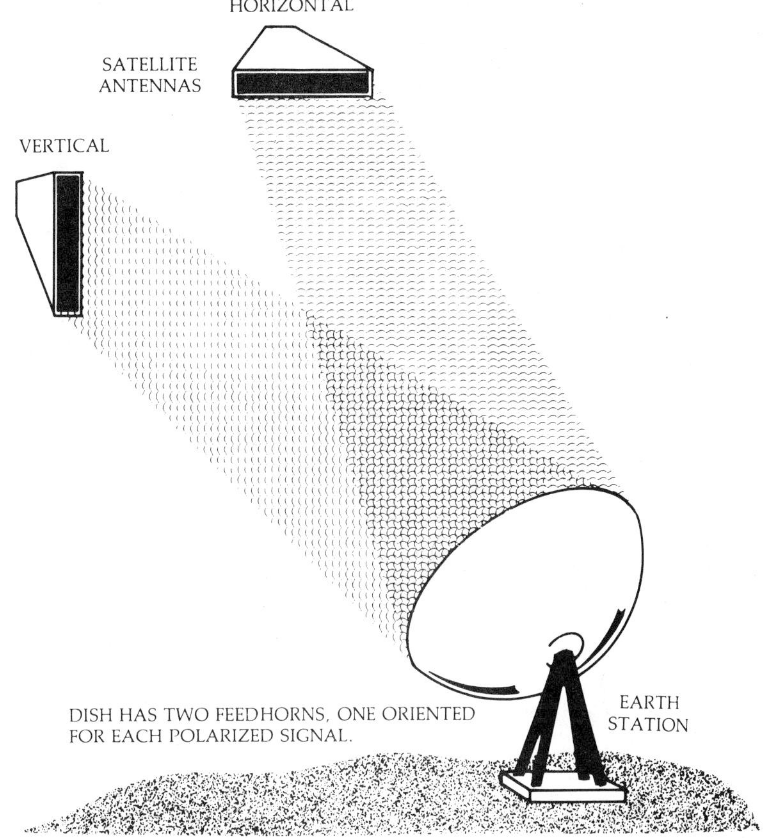

Polarization: How to double the "throughput" of a satellite transponder—send two signals at a time at the same frequency without interference, one polarized vertically and the other horizontally.

tally, TDMA (or some future improvement on it) will become practically universal.

POLARITY

It is also possible to use a single frequency *twice at the same time* through *polarization.* One analog message (or one TDMA cluster) can be sent through a "vertical" antenna, while another goes through a "horizontal" antenna. RCA demonstrated this technique in Alaska in the late 1970s, sending two different television programs simultaneously over a single transponder. Alternatively, an antenna can be configured to send signals "clockwise" or "counterclock-

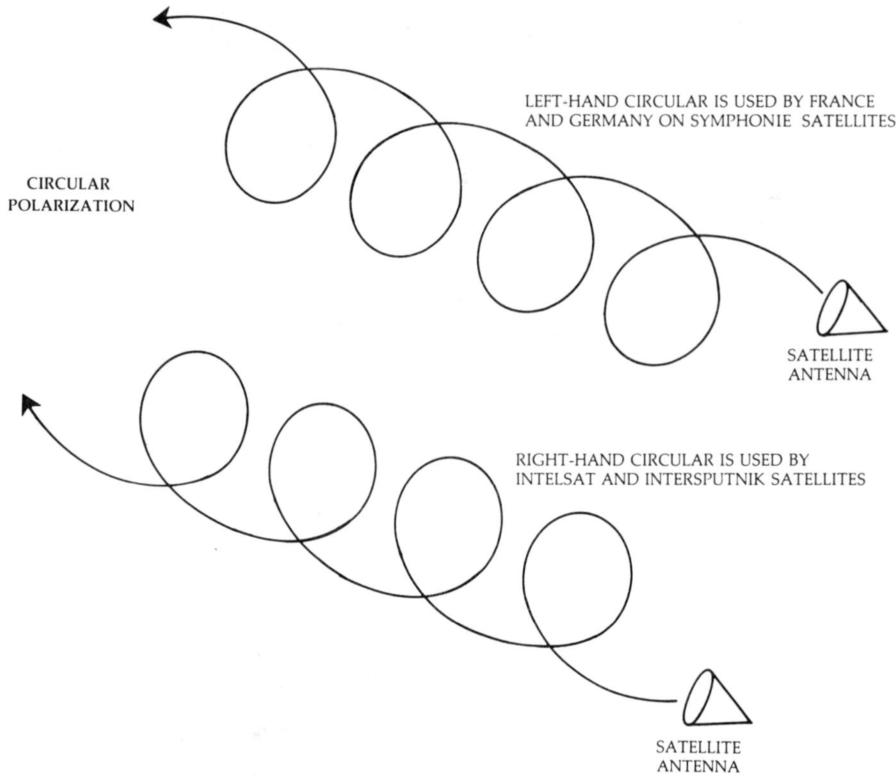

Circular polarization: Signals from a satellite transponder can be polarized in a circular pattern, either left hand or right hand.

wise'' using what is called *circular polarization.* Most terrestrial tv is horizontally polarized; that's why tv antennas are flat and parallel to the horizon. Circular polarization reduces the interference from nearby broadcast antennas. This two-for-one deal is remarkably free from trade-offs: it seems to be a genuine gain in efficiency without sacrificing anything else, except perhaps the money to install a second receiver and its hardware.

Endnotes

1. Frequency is expressed in cycles-per-second (hertz, abbreviated Hz), with metric prefixes for scale. Kilohertz (kHz) are thousands;megahertz (MHz) are millions; gigahertz (GHz) are billions. The actual numbers are these: a telephone call requires about 4000 Hz (4 kHz), and a color tv channel about 36,000,000 Hz (36 MHz). The mathematical quotient, 9000 phone calls per tv channel, has never yet been achieved in practice because the phone calls must be ''sep-

arated" from one another to avoid interference; it takes about 1000 Hz "between" them as a cushion. There are, however, some ways around those limitations, but they are technical. A simple explanation of two approaches is given in the section called *"Something For Nothing."*

2. An "uplink" is also the name for an earth station which can transmit to a satellite; and a "downlink" is one which can only receive.

Section 2

THE LAW OF THE SKY

"You'll always know your neighbor,
You'll always know your pal,
If you ever navigated
On the Erie Canal."
—Folk Song

Chapter 4

PARALLELS OF HISTORY

In the days when the words "United States of America" were still new, the young country wanted to grow westward. To haul pioneers from the eastern seaports to the interior farmlands—and fresh produce all the way back—the people dug canals.

Well, actually, only some people dug canals. The rest used the canals after they were in place, and thereby hangs a tale. Unlike public roads, canals were *private* avenues of commerce, and to ensure that they would be maintained in the public interest, the users made laws to govern them.

Letter carriers rode the paved highways, the canal boats and (later) the railroads. Lines of transportation and lines of communication were more or less the same; telegraph cables were later strung along those routes. Today, we think we can distinguish "transportation" from "communication" because their technologies are different; but socially, politically, and economically, the two are not easily separated. The legal framework that governs modern telecommunication networks was built by people who believed in the same ideals as did those who regulated the canals and railroads.

COMMON CARRIAGE

Serious intellectuals, fierce populists, land-hungry farmers, and adventuring speculators dominated the political life of the early United States. Yet all of their competing interests were served by easy access to canals.

What mail, newspapers, and decrees there were fanned out from the cities along the roads by horseback and by coach. Where the way was long, information rode the canal boats. The newly self-governing people—no longer colonial subjects—realized they were served best when they were free to use those canals, no matter who had dug them or who ran them. So the federal government, and that of the states through which the canals ran, required owners and operators of barge canals to allow *anyone* to use the facility, to make uniform rates for carrying freight and passengers, and to treat all customers equally, without discrimination against any individuals or classes of users. By accepting such regulation, the carrier gained security: no one would dig a competing canal.

When a carrier thus became a *common carrier,* it was charged by federal law to operate in the "public interest, convenience, and necessity." Specifically, that required it to publish *tariffs,* and to allow the government to approve its published rates according to standards of reasonableness and nondiscrimination. The common carrier could charge *only* those rates and no others. It was prohibited from abandoning routes, or discontinuing services, even when those activities were losing money. In the nineteenth century, the railroads were placed under the same rules.

In the nineteenth century, too, the dramatic advances in communication could have dissolved the bond with transportation, but they did not. Telegraph wires were capable of being strung where no railroad or canal could ever go, especially over the western mountains but, by and large, they continued to follow the existing trade routes. Older technology disappeared swiftly: stage coaches no longer found business in any town that got a railroad station, and the pony express withered away while its fastest riders were still young men. Yet the rules of common carriage were almost immediately applied to telegraphy, and for much the same reasons: laying cables was an expensive proposition, and no one wanted to invest in it if there were too much competition. To preserve what was later called their "natural monopoly" in communication, telegraph companies agreed to be regulated as the canals and rails had been.

THE EUROPEAN ALTERNATIVE

After 1812, and with the exception of the short Franco-Prussian War, Europe experienced a rare century of almost complete peace. International commerce and cooperation grew. It was essential for the success of telegraphy. Yet, Europe was a continent-size Babel where everyone of importance spoke a different language. Its experimental scientists had no professional journals through which they could share ideas. Books were not published or translated until years after a piece of work was completed. When the first International Electrical Congress met in 1881, there existed 12 different units of electromotive force, 10 different units of electric current, and 15 different units of resistance. In effect, the congress adopted Ohm's "law" fixing the relationship of what we now call the ohm, volt, and ampere to one another.

The International Telegraphic Union was formed in 1865; without it, there might never have been uniform message services or a transatlantic cable. The telegraphic union started calling itself the International Telecommunication Union (ITU) in the 1920s, and is now a part of the U.N., but it remains the oldest continuous international deliberative body in the world. Several other overlapping organizations now set technical standards for electrical and telecommunication equipment. This achievement—international cooperation to share and nurture a common technology—was the first "common market" that Europe had ever known, and surely the first in the world that was not the fruit of conquest.

But European countries, and their governments, were far more centralized than was the American experiment. Despite the "modern" idea of international cooperation that produced the ITU, the internal affairs of each country tended to be handled in very traditional ways. Centuries of monarchy had vested control over public works and other services in well-entrenched bureaucracies. Even where constitutions and parliaments had replaced the arbitrary whims of hereditary rulers, the day-to-day power over roads, rails, and mails was held by civil servants. Of those services, almost without exception, the post offices were the most vital. They transported the documents of commerce and government.

In general, European post offices gained control over telegraphy. The technology—though faster—seemed to resemble the system of mails, and especially in small countries it was convenient to have agents already in place who could send and receive messages. Also, unless a recipient were a telegrapher, or worked in the telegraph office, messages had to be physically transported to that person's home or place of business, so telegraph messengers were recruited just as mail carriers were.

To this day, European countries and their former colonies—almost without exception—have post-and-telegraph monopolies that are either government agencies or quasi-independent corporations answerable only to the government. Only in the U.S.—and to a limited extent the U.K. and Australia—is a telephone or telegraph company owned by common stockholders. Having civil servants run the telecommunications gives users a quirky balance of services. Post-and-telegraph monopolies boast of enormous economies of scale, uniform technical standards, and one-stop communication "shopping," yet in many of those countries, the waiting list for ordinary phone service is nearly as large as the telephone directory itself. Nonetheless, where it is established, the local post-and-telegraph organization is the one charged with exploring or operating satellite communication networks.

THE PUBLIC INTEREST, CONVENIENCE, AND NECESSITY

Americans have a broad distrust of monopolies, which helped them maintain the subtle regulatory link between communication and transportation well into the twentieth century—even to the present day. From 1910 to 1934, American common carriers were regulated by the Interstate Commerce Commission (ICC). The right of the federal government to control transportation, even when it was wholly within a single state, was established at the turn of the twentieth century. The U.S. Supreme Court's decision in the case of Houston, E. & W. T. R. Co. vs. U.S.—better known as the "Shreveport doctrine"—held that an intrastate railway was part of a nationwide network of rail service, and therefore federal authority superseded that of the state of Louisiana.

That precedent was carried over into communication in 1934, when the Senate Commerce Committee—too busy to draft an original communications law—"borrowed" some attorneys from the ICC. Together, they wrote what became the 1934 Communications Act, incorporating the 1927 Radio Act that had dealt with broadcasting.

The 1934 Communications Act economizes a scarce resource. It is predicated on several notions, related to a common theme: that there are only so many radio frequencies, so the government must act in the public interest to assure everyone an equal right to use them. As a result of that axiom, the government regulated and policed the frequencies assigned to broadcasters, and required carriers to file tariffs. Frequencies were *assigned* to licensed operators for a few years at a time—not granted or sold in perpetuity. Many of its rules, such as tariff procedures, were copied from the interstate commerce laws not only out of convenience but because they represented the philosophy by which the U.S. intended to regulate domestic telecommunication: that is, a philosophy of equal, nondiscriminatory access. That philosophy was similar to the one which had established a strong federal government to mediate among 13 quarrelsome ex-colonies: that only government, *a representative government,* could ensure that these very limited resources were equitably parceled out.

The chief requirement imposed on an American common carrier is that it should operate a communication channel only as long as it does so in the "public interest, convenience, and necessity." The reality, however, has been that once a person or organization is licensed to operate a communication channel, it is nearly impossible to oust them. The U.S. Department of Justice tried for generations to overcome the concentrating tendency at AT&T, but only in 1982 did it obtain a judge's ruling that the telephone giant be divided into smaller, autonomous companies. During World War II, the U.S. government monitored transatlantic telegraph cables between Europe and South America, and found that the Germans were using them to send messages to their agents in the western hemisphere. The government did not sever the cables; nor did it prosecute the owners (U.S. companies) or lift their licenses for carrying enemy traffic. Instead, it continued to wiretap the cables, and returned full control to the carriers after the war.

Although television and radio stations are not, by definition, common carriers, the FCC has always required them to show concern for the "public interest, convenience, and necessity," before their licenses can be renewed. Yet—like common carriers—few have ever lost a license. The FCC lifted RKO's Boston tv license in 1982 for corporate misconduct on the part of RKO's owners, but the FCC has revoked practically no other tv licenses recently. It did lift a handful of radio licenses, in the 1960s and '70s in some southern states, as punishment for broadcasting grossly heinous racial slurs. Lord Thompson of Fleet, the Canadian/British media magnate, has been quoted as saying that owning an American television station was like having "a license to print money."

MASS MEDIA

When the Erie Canal was dug, it was the only way to get from the Great Lakes to New York sitting down. But it was just so wide, it could hold just so many boats in a lock at one time, and no more.

Regulation maximized its traffic, and kept it running efficiently. Competition from railroads—not government regulation—ended the useful life of canals; later, in the 1950s, competition from highways and airlines drove some railroads into bankruptcy.

There were no competing communications technologies in the 1930s. The 1934 Communications Act was designed to keep communication channels out of the hands of monopolies, although it also legitimized some de facto monopolies. It allowed AT&T to dominate the telephone business because "universal service" was seen as more important than competition; and the act perpetuated the role of the radio (and later television) "networks" long after technology had developed other options for distributing programs. Most adults had living memory of robber-barons and their "trusts" that had gained control over whole industries, such as oil and steel. If they had lived to see the hodgepodge that actually resulted, the act's authors might be appalled, but, of course, they had no other way of interpreting history or predicting the future. Besides, at the time, communication still did not seem to be any different from transportation.

Information has always traveled by whatever means it could find, and the faster the better for people who need it. The telegraph and, later, the telephone were accepted immediately wherever they were installed. The nation was hungry for news. Abraham Lincoln lost the Illinois senate race in 1858, but he "won" the debate against Stephen Douglas because reporters (using the new "shorthand") sent the entire transcript to every newspaper that had a telegraph station; Lincoln became, in effect, the first politician to command the attention of the American electronic media.

Great, innovative ideas often appear simultaneously in a few contemporary minds: calculus (Newton and Leibniz), natural selection (Darwin and Russell), and television (Zworykin, Farnsworth, and others), to name just a few. But one man created the modern structure for communication: David Sarnoff invented *broadcasting* as we know it.

There had been occasional radio broadcasts even before World War I; some shipboard wireless operators thought they were hearing the "music of the spheres" through their earphones, but what they really heard were amateurs' experiments with musicians and antennas. Thomas Edison himself did not think radio broadcasting would catch on, but—being nearly deaf—he could not really understand it the way other people did. Sarnoff understood communication deeply, probably because he had been a wireless telegrapher himself. One fateful night, the Marconi station he was working in served as New York's connection with the foundering R.M.S. *Titanic,* and Sarnoff became a hero.

During the years Sarnoff ran RCA, broadcasting became a multimillion- (later a multibillion-) dollar business. In large part, that was because the 1934 Act held to the doctrine of regulating scarcity. There were only so many frequencies on which to transmit radio programs, so the federal government parceled them out to applicants who appeared likely to run them in the public interest. From the first, these tended to be very rich men and their corporations, particularly those that owned newspapers, magazines, theaters, film studios, and other communication resources.

Local stations stayed "local" until Sarnoff put together a *network* of them. He leased special telephone lines from the Bell system and transmitted identical programs through them to those "local" radio stations all over the country. In the 1920s, radio was a novelty; by the end of World War II it rivaled newspapers for the public attention, and by the 1960s, television news reached more people than newspapers did.

As the business of broadcasting grew, though, it assumed many of the characteristics of a monopoly. The American Telephone and Telegraph Company had long since persuaded the FCC that a "natural monopoly" should exist in telephone service. Broadcasters developed a network of stations that were "affiliates," if they were not entirely "owned and operated" from a single corporate headquarters. During that time, the protection of freedom of the press came to reside mainly in those who could afford a printing plant, and responsibility for ensuring freedom of speech was increasingly vested in the stewards of the airwaves.

THE POLITICS OF SCARCITY

Radio waves are electrical currents that can be induced in antennas by a radio transmitter. The "waves" or electrical vibrations "wiggle" their way up the wire or pole and out the end. Similar to light from a candle, radio energy is uniform in all directions, so much of the energy escapes into space. Receiving antennas that are sensitive (tuned) to those waves can be made to oscillate just as the transmitting antenna did. The rate of oscillation—the number of times per second that they make complete cycles—is called *frequency.* There is a very wide range of potential frequencies over which radio waves travel, but the number of *usable* frequencies is ultimately limited by people's ability to propagate them. As an analogy, consider that the actual colors in a rainbow are subtle and uncountable, but shades of paint are not.

Cooperation among the members of the ITU and other organizations has always been based on the assumption that the electromagnetic radio "spectrum" is limited. Once the technology was available to make rockets, the concept of scarcity was extended to outer space, imposing a limit on the number of places high above the earth where it is possible to place communication satellites (see Chapter 4).

In the ITU, international conventions, treaties, and unspoken agreements all hold that the spectrum of usable radio frequencies and the number of potential satellite orbit "slots" are limited natural resources that have to be conserved—or, at least, wisely managed—by mutual consent for the common good. The record of its first 100 years is one of almost unbelievable cooperation, even during wartime. Except for "jamming," practically no nation has ever broken the ITU's agreements. Sometimes, nations have transcended their own pride and politics to reach consensus with others.

But in the last quarter of the twentieth century, technology has opened up many more spectrum "bands" than ever before, and new pressures have been exerted that would increase the number of satellites aloft. Cooperation has diminished, and the more powerful countries have pursued their own interests at the expense of the weaker. Unfortunately, the U.S. is widely regarded in the world as the chief offender. It is home to many transnational communications corporations, and does not regulate their behavior according to the wishes of other countries in which those corporations do business. Americans are the world's leading suppliers of films and television programs which influence public opinion in many developing countries; and which also compete for scarce broadcasting time and money. Being able to build and launch virtually any satellite which its people would want, the U.S. has defined the state-of-the-art in terms of its own capabilities, and wants international agreements to conform to them after the fact.

The U.S. has also begun to sever the last connection between transportation and communication. The official attitude of the U.S. government toward the propagation of radio waves and the launching of satellites is that there is an *abundance* of communication resources. The policy which that attitude engenders was summed up in

1982 by William Baxter, assistant attorney general for antitrust. The radio spectrum and the satellite orbit are now classified as real property, to which the U.S. wants to apply the same philosophy by which it has tended to manage its abundant lands, minerals, and timber. Baxter said this: "There is no spectrum scarcity. That's a myth. We should treat the electromagnetic spectrum just like the surface of the earth. People should be able to own it, sell it, and cut it up into pieces."

That is a fundamental challenge to the way the rest of the world sees it. The issues are far-reaching, and ultimately affect everyone who has (or hopes to have) a telephone, a television, or a computer.

Chapter 5

SPACE LAW

When passengers can ride the space shuttles, one of the first will probably be a lawyer. Even now, outer space has been so effectively colonized that a body of law has grown up to serve it.

Everyone who read science fiction in the years following World War II felt that space travel was not only possible but inevitable. Science teachers, engineers, and physicists who reveled in Sputnik, Telstar, and the moon landings were right at home, but people in the legal professions seemed strangely out of place in that realm until recently.

An Outer Space Treaty, negotiated through the U.N. in 1967, declared space open for exploration by all states without discrimination; celestial bodies were specifically protected from national appropriation. The Treaty's legislative pronouncements, however, are too broad to address narrow details, and many of its ideas are subject to interpretation.

The astronauts who landed there declared the Moon to be the common heritage of all mankind, but the image the world saw on tv was the planting of the American flag. Whose moon is it? Three Soviet cosmonauts lived in a "space station" for 211 days in 1982, where they were probably exploring military strategies for outer space. Was that a violation of the treaty?

Day-to-day issues have begun to clog the agendas of international debate—issues such as: which nations, corporations, or institutions may use or launch a satellite? Where shall its orbit slot be allocated? Over what frequencies shall it receive and transmit? Who is liable for the content of what is broadcast?

Some questions have been shelved: one is whether air-rights of a country extend into outer space, as some equatorial nations have claimed; the consensus in the U.N. and the ITU is that they do not. But other questions nag at the world's conscience. Does a country have the right to use—or reject—signals falling onto it from a neighboring country's domestic satellite? Shall outer space belong to those who can afford to use it, on a first-come, first-served basis? Or should some orbit slots and frequencies be held open or pre-assigned for the developing countries, so they can grow into them?

The central problem of space law is that there is no clear precedent for it. The new Law of the Sea copes with many similar questions, but as a model law it may be inadequate; exploitation of space is not as well developed as that of the oceans, and there is no equivalent to the "coastal" zones over which maritime nations have traditional claims. Moreover, in several nation states, the U.S. among them, domestic and international policies are responsive to and shaped by many different interest groups.

THE BODY OF LAW

Treaties are signed by members of the United Nations who are willing to compromise, trust each other, and hope for the best. But that spirit of cooperation arises only from a foundation of security that one's own interests have been protected.

Neil Hosenball, general counsel for NASA, says that four treaties are now in force that affect international satellite communication. He prefers to use their common names, rather than the long and formal titles they carry.

> The first was the 1967 treaty of Outer Space Principles [mentioned previously] which is a kind of 'constitution,' spelling out the broad principles. The second, which was ratified a year later, is the Astronaut and Space Object Rescue and Return agreement. That's the document that specifies and carries out the intent of the 1967 treaty.
>
> The Outer Space Liability Convention, from 1973, set up an international claims commission to handle problems with collisions in space as well as injury on the ground from falling debris.
>
> Then, in 1974, we got the Registration Convention, under which all countries report their launches to a U.N. registry.
>
> There was a fifth one, known as the Moon Treaty, that

emerged in 1979, but that has never been signed by more than 11 countries; the U.S., the U.S.S.R. and the others were put off for the same reasons that later drove them to reject the Law of the Sea treaty—it tries to bring the Moon into the "common heritage of mankind." But the Moon Treaty mainly said that, when exploitation of the Moon became feasible, there should be another convention to decide how to do it.

Besides those treaties, there are the agreements which form the ITU, INTELSAT and INMARSAT. "With them, you have international space law as we know it," says Hosenball.

The U.S. has its own body of law regarding outer space. That begins with the 1934 Communications Act which created the FCC and gave it the power to regulate domestic communication. In 1958 we got the Space Act, which created NASA, and the FAA Act, which gave the FAA jurisdiction over rockets, balloons, missiles and other things that fly through American air space. It was under that Act, by the way, in 1982, that the FAA prosecuted a fellow who flew in a lawn chair with helium balloons.

In 1964, Congress passed the COMSAT Act, which established COMSAT as the sole international satellite common carrier.

There is also a section of Title 18, in the federal code, which extends the concept of extraterritoriality to space vehicles—in case the government ever has to pursue a criminal beyond the earth. And there is an Export Control Act provision that gives the State Department control over private, non-governmental rocket launches into space—like the one from Texas in 1982.

THE INTERNATIONAL DEBATE: CIVIL LAW VS. COMMON LAW

The international debates are the most complex but, seen in a certain light, they may be more easily understood than may the domestic ones. According to Edward W. Ploman, vice-rector for global learning at United Nations University, in Tokyo, the differences among countries' attitudes toward space are rooted in their different concepts of law itself: "civil law" or "common law," Ploman says the differences "point to almost opposite ways of looking at the role of the legislator and the intent of legislation.

The civil law tradition originated in Roman law and subsequently entered Islam. "The legislator," Ploman says, "creates order out of chaos. It is therefore not only logical but also imperative that the legislator should act as early as possible and provide rules for all foreseeable cases."

> In contrast, the common law approach has been described as assuming a pre-existing "living" law, which has not necessarily been codified and which it is the task of the judge to apply in concrete cases. The intervention of the legislator should therefore ideally only be necessary to fill the gaps, as it were. There follows a distrust of generalized, over-all, systematic approaches, which are essential for the civil law tradition, and a preference for legislation as late as possible and then only in very general terms.
>
> A common-law-inspired attitude would then naturally wish to avoid any satellite broadcast principles until the technology was fully developed and there was a proven need for such rules. This position would, from a civil law approach, be interpreted in a negative way as prevarication or worse. The common law countries would see the other attitude as equally awkward in terms of a rigid, seemingly artificial approach which might foreclose possibilities and hinder development.

In this analogy, the U.S., the U.K., most of the Common Market, plus the U.S.S.R. and the Warsaw Pact countries take a common law position in international debates over space. Theirs is an after-the-fact approach (called *a posteriori*) holding, essentially, that once a thing is done it should be accepted as fact, and that subsequent regulations—if any—should not change that fact. It is, in other words, squatters' rights. Countries which view outer space in this light support its free exploration, and insist that its resources should be exploited by whomever is able to do so. Recent American domestic ventures into "de-regulation" come from that attitude, which is boosted there by an underlying philosophy of *laissez-faire,* and grounded in a Constitutional provision which says that rights not specifically granted to the federal government belong to the states or to the people themselves.

While the Soviet Union and its socialist allies are not so free, politically, their legal systems—like those of the West—come from a European philosophical tradition in which entities such as "state," "citizen," etc., are accepted as real things, having *contractual* relationships with one another. Originally framed by scholars of mathematics and physics, social contract law is the "natural law" or (in Ploman's words) "living law" which can be cited as underlying the common law tradition.

The nations of Asia, Latin America, the Middle East and Africa, on the other hand, tend to take the civil law approach. Ploman says that is because their cultures are derived from different principles. The *sharia* of the Koran, and the *dharma* of the Hindus, for example, are frameworks which encompass the whole of human life, temporal and spiritual, not merely the "laws" of governments or the "rights" of citizens. Recourse to legal codes or judgments is often seen, in that light, as a last resort after mediation has failed.

Since these countries feel they have been unsuccessful at getting the industrial nations to act on their concerns, they now argue for allocation of orbit slots and frequency assignments in advance of demonstrable need. The developing countries feel that there may always be a discrepancy between their own technologies and those of the industrial nations, and that they will probably not be able to exploit outer space resources as quickly as the advanced nations have, however desirable such a venture might be for domestic or regional development. So they ask for allocations prior to need (*a priori*).

Fear of being left behind is one serious motive, but there are others. The "Southern hemisphere" nations are joined by Canada in being keenly sensitive to what has been loosely called "cultural imperialism"—the domination of domestic broadcasting by foreign mass media, most of which is owned or controlled by people in the "Northern hemisphere," home of the industrial countries. They have therefore tried to impose "guidelines" on reporters and on the flow of news, or to establish sanctions in the form of laws governing trade in information.

(In Canada's case, says Ploman, the "overwhelming presence south of their border makes Canadians question their national identity, and keeps them more aware of issues that affect them as a nation. They have thought about telecommunication problems, perhaps more than anyone else.")

In their colonial periods, European nations introduced their legal systems to the colonies. France imposed the civil law tradition in

North America, West Africa, and Indochina. Britain imposed the common law tradition in India and Nigeria. When the colonies gained independence, some retained the colonial legal outlook and others did not. Some countries, such as Japan, adopted forms of both traditions, accepting the German idea of a constitution and a legislature, while retaining the older Chinese influence in not wanting to go to court over disputes, and relying on mediation and consensus.

Because of the diversity, "perceptions of right and wrong are as much a part of a national 'reality' as anything else is. You have to take them into account when dealing with other countries," says Ploman, "but the western nations—especially the U.S.—often refuse to consider them. If you don't know what people are really talking about, you tend to think of their complaints as *rhetorical* concerns, instead of real problems. When you treat those problems only as rhetorical, you fail to deal with *real* issues."

For example, developing countries which call for a "new information order" cite abuses by the industrial countries' media with words like "aggression," "imperialism," and "neo-colonialism." Ploman says people who use these words feel vulnerable and concerned that new technologies may be harmful to their national interests. "The West should show more concern for opinions that are *removed* from their own."

The U.N., the ITU, and other international deliberative bodies are the creations of Western and Westernized legal minds, says Ploman. Their treaties and decisions "contain such expressions as 'freedom,' 'rights,' and similar abstract nouns without a precise definition and therefore without exact normative content." All the countries, therefore, are free to interpret them as they choose, and so the cycles of argument and counter-argument go on.

THE LAW OF THE SEA

The Law of the Sea treaty may be a model for the international debates of the future. In the 1980s, the number of countries mounting maritime activities is growing: many small countries now have not only navies but research ships, factory ships, fishing fleets, and oil drilling platforms. The cost of these ventures is high, but not prohibitively so, and the risks of depletion and pollution have never been greater.

Elizabeth Mann Borgese, a political science professor at Dalhousie University, Nova Scotia, watched the debates over the Law of the Sea for many years. They finally produced a Convention, in 1982, which divided most of the world's oceans into fishing zones, economic zones, and territorial zones; it also deeded about 40 percent of the oceans, particularly their mineral resources, to the "Common Heritage of Mankind."

"The Convention," she says, "clearly exhibits the shaping by parallel yet often intersecting revolutions in technology and international relations. It was the technological revolution that pushed from the North the expansion of the national jurisdiction in ocean space. It was the revolution in international relations that pressed, from the South, for the establishment of new, strong international institutions." Replace the words "ocean space" with "outer space" and her perspective applies to space law.

As with the resources of outer space, the Southern hemisphere nations wished to begin exploitation in advance of need, with a worldwide Enterprise organization that would compel Northern nations to provide the technology. The industrial countries preferred to let their own national and transnational organizations exploit the ocean's resources at will, but were prepared to pay a tax or a license fee to the Common Heritage fund based on whatever they extracted. A compromise suggested by Henry Kissinger, says Prof. Borgese, "combined the impractical Enterprise with the unacceptable licensing agency in a 'parallel system,' a combination which, by a kind of witches' arithmetic, was to result in a solution both practical and acceptable: The Conference never quite recovered from it." Unsatisfied with the outcome, the U.S., the U.S.S.R., and 15 other nations refused to sign the Convention.

Will the ITU suffer from the same kind of paralysis? Richard Butler thinks that it will not; he is its Secretary-General. "It doesn't take long for purely national communications to become international, and the U.S. understands that very well. The pressures for standards are too great," he says, and points to successful cooperative ventures in the ITU that led to worldwide agreement on distress signal frequencies (see Chapter 15). "Unlike the Law of the Sea treaties," he says, "the ITU has provisions for a country to record 'reservations' about a document, which can be either provisional or final. Total absence would deprive a country of the right to make those reservations."

NEW ACTORS ON STAGE

Edward Ploman says that the reason why the U.S. did not recognize that other countries were so concerned about prior allocation of ocean resources was because transnational interests had complicated the picture. "Fishery people, seabed mining people, and so on, had combined, internationally, according to their common interests, and occasionally took sides against their own countries' policies!"

Ploman expresses concern about those entities he calls "actors" on the stage of international affairs and deliberation, not only in the Law of the Sea but in other areas, especially communication. Those entities, he says, are not governments, but those which form "criss-crossing patterns of influence and lobbying at the international level." Because, as he says, "nuclear power, space power and information power tend to be concentrated in the same states, 'sovereignty' is no longer accepted as absolute and indivisible, but as a relative concept that encompasses a great variety of national sectors on the international scene, from superpowers to mini-states, with some being definitely more equal than others."

Nation-statehood is no longer a prerequisite for participation, nor is it the highest expression of a point of view, so the power of nonstate interest groups has grown, and that worries Ploman. "The optimists tend to view these developments as the emergence of new communities of interest made possible through new communication technologies. In a more academic vein, analysis is made of 'epistemic' or 'cognitive' communities. But one could also talk of pressure groups, or simply of international mafias. Simply to call them 'nice communities of interest' is not enough. Many of them act like international mafia."

DOMESTIC ISSUES: AMERICAN DILEMMAS

Is the Law of the Sea Convention a precedent for outer space debates? That will depend on how well the industrial nations manage the conflicting pressures from the "actors" in their domestic telecommunication debates.

The policy of the U.S. toward space is called "open entry" or "open skies." If an applicant wants an orbit slot, and can be reasonably expected to fill it, and there is no compelling reason for denying it, the FCC will probably grant that applicant permission to install a satellite there. But if two or more applicants want to use the same slot, there is supposed to be a hearing.

"You have to have comparative hearings to decide which is the better applicant," explains Veronica Ahern, a communications attorney who formerly worked for the State Department. "The Supreme Court took that position in 1945, in the case of Ashbacker Radio Corp. vs. the FCC. Now, the expression 'Ashbacker doctrine' is applied to proceedings that review mutually exclusive applications and select the most qualified one."

"But the FCC has avoided Ashbacker-type proceedings in the area of domestic satellites," says Ahern, "because they are long, costly and mainly profitable only to lawyers. The Commission has held 'rulemaking' proceedings instead, to see if any of the applicants, or whole classes of applicants, may be ineligible." Auctions and lotteries have been suggested, and gained public attention, and one was held at Southeby's, the famous auction house. The "winners," however, were disappointed when the FCC later invalidated the auction and required the satellite operators to come up with a more equitable scheme.

"Until recently," says Ahern, "the FCC had the flexibility to move satellites around to make room for new ones. Now that the available arc's most desirable slots are filled, it's conceivable that the Commission will have to undertake Ashbacker-type findings. If company A and company B both want the same slot, the FCC will have to find a way to decide between them. The issue of the 1980s will be: what are the criteria the applicants will have to meet before the FCC grants them a slot for satellite service?"

"Members of the industry *could* adjust their own requirements without risking antitrust violations," says Ahern.

> 'It happened recently in the case of cellular radio. There, applicants for intra-city paging and mobile telephone frequencies were told in advance that there would be two licensees in each city: one for the local telephone company and one for an independent company. In nearly every city, the independent applicants pooled their proposals in some way. Several actually merged or were acquired by others, leading to suggestions that they had entered the proceedings simply to speculate in the value of their applications, and fully expected to be bought out. Nonetheless, the awarding of cellular radio licensees has proceeded very smoothly.'

"From an American consumer's point of view," Ahern says, referring mainly to business users, "it's been a very successful decade for satellite services. Demand has been a sort of self-fulfilling prophesy, as increased use of domestic satellites for cable and network TV program distribution has stimulated the design of flexible systems capable of handling those uses. Southern Pacific, MCI and other competing long-distance services are proposing to have their own satellites."

The ease with which satellites are placed in orbit may be restrained somewhat in 1985 when a scheduled World Administrative Radio Conference deals with issues of space and satellites. "The 'Space WARC' will probably guarantee access to orbit slots for all countries," Ahern predicts, "and that will require *a priori* planning. Geostationary orbit positions, and the frequencies over which satellites in those slots communicate, may be allocated—not on a basis of need, but—on the stated desire of countries to have their fair share."

RIGHTS OF ACCESS

Public *access* to a common carrier is at the heart of American communications law. Americans have come to expect a "universal" telephone service. That end is clearly socially beneficial, but achieving it requires cross-subsidies: the transfer of revenues from high-profit services to cover those which incur deficits. (INTELSAT has also found it desirable to cross-subsidize operations in the same way.) Public utility commissions and other local regulatory bodies set limits mainly on the overall profitability of a telephone company and require only that local rates be kept as low as possible and that everyone in the service district of the carrier be afforded "basic" service, i.e., a telephone.

Two recent changes in local telephone services bear on the role of satellites: de-regulation of "customer premises equipment" (CPE) and the imposition of fees for access to telephone networks by outside carriers.

CPE is whatever a customer has on the premises; some equipment is supplied by the phone company, while other equipment may be leased or purchased independently. To residential customers, CPE is fancy or exotic telephones in new shapes, and perhaps an answering machine. To business customers, CPE includes private branch exchanges (PBXs) and switchboards, computer modems [for passing computer data along phone lines], and other signaling, communicating or storage devices. The Supreme Court's "Carterfone" decision of the 1960s established the principle that, as long as a piece of equipment met minimum standards of safety and electrical compatibility, the telephone company could not prohibit its connection into the system.

By the beginning of 1983, the FCC had de-regulated CPE in such a way that if a customer did not lease CPE from the phone company, that company could not charge a fee in lieu of such a lease. Essentially, the customer was freed to install whatever CPE met his or her needs.

"For the FCC to do that," says Veronica Ahern, "it had to assert its jurisdiction over CPEs, which had been under state PUC jurisdiction up to that point." To do so, the FCC relied on the "Shreveport doctrine" (see Chapter 4) which had subjected intrastate railroads to interstate regulation. "The FCC said that CPEs performed both intrastate and interstate functions and, in effect, said that 'our way of regulating them will be to de-regulate them.' The National Association of [State] Regulatory Utility Commissioners appealed the ruling to the U.S. Court of Appeals, but a decision has not yet been handed down."

Access charges are imposed whenever an outside carrier wants to patch into a telephone company. Sprint, MCI, the newly deregulated AT&T, and the other independent long-distance carriers, for example, pay such a fee in each city or service area in which they operate. Those fees are typically set by local PUCs. "Where commissioners are elected by the voters," says Ahern, "they usually run on platforms of keeping service rates down, or of getting rebates from utilities to customers. The junior senator from Florida, Paula Hawkins, was a PUC commissioner whose popularity came from getting rebates."

Commissioners set access charges after consideration of the phone company's allowable rate of return on investment. "If it is the commissioners' wish to keep local 'basic phone service' as low-priced as possible, one way is to grant the phone company the right to charge higher access fees to outside carriers. That subsidizes local service at the expense of outsiders. The tightrope they are walking," says Ahern, "is that if access charges go up to an unacceptable level, the carriers have a financial incentive to *bypass* the local phone company altogether and go directly into the customer's premises." *That* falls under the heading of CPE and is therefore not regulated!

Such CPEs might include microwave networks, VHF radios, and even infrared and visible-light laser systems (which, by the way, are not regulated at all). Where there is cable tv installed, access to those lines may be cheaper than access to the phone lines. And for large customers, or for those in a single geographic region, there is the possibility of having their own satellite dishes.

Direct-broadcast satellites, beyond their entertainment potential, can carry intercity paging, computer data services—the other "public" communications that have been the province of telephone companies and specialized common carriers until now. "It's unclear," says Ahern, "whether the state PUCs or the FCC will ultimately have jurisdiction, but I think that the access charge problem is a nationwide one. If the FCC's policies remain consistent with the direction they have been going—like those of the federal government as a whole—then it will be the FCC that makes the rules."

RESERVATIONS FOR ORBITAL SLOTS

The U.S. is clearly the world leader in telecommunication technology, but it may not be able to keep the political lead which it has enjoyed. It maintains a dualistic domestic policy that tries to respect local marketplace decisions while still relying on a strong federal government to serve as judge-advocate. Those ideas may not be acceptable to the rest of the world.

Since orbit slots and frequency spectrum bands are phenomena of nature, and can be regarded as natural resources in that way, then what historian Barry Commoner calls "the tragedy of the commons" is an instructive allegory. Consider a town with a large, grassy, public field—called "the commons" in Europe and New England. As more and more shepherds graze their flocks in the field, the grass cannot regenerate; none of them owns the field, and so none takes responsibility for its continuity as a source of grass. Yet, *because* none of them owns it, no one has authority to keep others out—even temporarily, while the grass reseeds itself. Furthermore, none wants to take his sheep out of the commons, because that would leave "more" for the others; each fears that the rest will gain at his expense if he voluntarily withdraws, so the incentive a shepherd feels is to graze more—not less.

Developing countries want access to orbit slots and to the frequencies over which satellites in those slots can transmit and receive signals. Historically, they have found themselves frozen out of some desirable slots and spectral bands because industrialized countries got there first. The problem is that there are only so many "good" slots—those vantage points from which an entire continent, or its most affluent region, is under the footprint. If everyone tried to put satellites into those slots, the radio signals would interfere with one another, limiting everybody's access. So the ITU allocates orbit slots and spectra worldwide, in a manner similar to the way the FCC allocates frequencies in the U.S.

If less-developed countries (LDCs) obtain orbit slots before they can use them, then the remaining slots become scarce; efficient technology for filling the slots grows more expensive—thus favoring the industrial nations who can still afford to fill them. If the LDCs are denied orbit slots, then they lose the chance to use current technology in the future when it will have become cheaper. Stuck in this no-win situation, the LDCs nonetheless want *a priori* (in advance) allocation of orbit slots and spectra for the same reason that shepherds want to graze the commons as much as they can.

The developed countries, and the transnational corporations which are headquartered in them, are responsible for most of the research and development, applications and products that exploit the common resources of orbit slots and spectra. It is unreasonable to expect them to do so without some reward in the form of "rights" to their inventions. Take just one classic example: dozens of companies made and sold telephones in the first few years of the twentieth century, but the Bell System argued (successfully) before the Interstate Commerce Commission that it could not develop a nationwide network without uniform equipment of its own design and manufacture. By becoming a "natural monopoly," which was consistent with the technology of the time, and probably in the best public interest at that time, Bell gained so many customers that it could actually *reduce* its tariffs periodically; they became so low that—for 60 years—no serious competition was possible. The users clearly benefited, and the company became the largest in the world. The only losers, at least at first, were the independent telephone makers.

ALTERNATIVE WAYS TO ALLOCATE ORBIT SLOTS

Despite several international brainstorming sessions, called World Administrative Radio Conferences (WARC), allocation of orbit slots has proceeded on a first-come-first-served basis, with the best ones taken mostly at will by the developed, industrial countries, and the WARC delegates more or less acceding to the *status quo.*

According to William Melody, of Simon Fraser University, less-developed countries could band together and force the industrial ones to accept *a priori* allocations, but that would generate resentment at least as strong as the backlash against the Organization of Petroleum Exporting Countries, and could endanger many ongoing financial and cultural exchanges.

Meheroo Jussawalla, a University of Hawaii professor, says flatly that "the third world countries have been denied the right to use spectra and orbit slots by their lack of funding. Yet even now, when some of them can launch their own rockets [as India does], they may not want to hurt their relations with developed countries."

Richard Butler, secretary-general of the ITU, says that "planning" is the most successful approach to spectrum and orbit slot allocation. "It was successful, after World War II, in the creation of a worldwide aeronautical mobile radio service, so pilots everywhere could use compatible equipment. National planning has *always* been carried out after the international decisions have been agreed to. Allocation of broadcast services, for example, has been characterized by fixed allocations and specific limits of service areas. Now, of course, countries want all the new services without having to pay a penalty for being late."

It requires action by the whole community to solve the problem of the commons, wherever it is. Harvey J. Levin, of Hofstra University, New York, has developed what he calls "five scenarios to reconcile divergent orbit and spectrum claims of rich and poor nations." In all of them, the inherent value of the orbital slot or the spectrum band associated with it can be "recovered" by the country to which it is assigned.

The Credible Guarantor

"The world's leading spectrum users would essentially agree to 'find' orbit slots and spectrum for latecomers when the latter's technology, know-how and domestic community needs had advanced to the point where they themselves could establish new domestic satellite systems. . . . To make this option sufficiently credible . . . the guaranteeing nations' seriousness of intent must be established by such initiatives as pledges to phase out or re-locate satellite operations upon request (by the LDC) in light of a demonstrable technical-economic plan and timetable."

Levin also raises the chief problem with this approach: whether the advanced nations can make really *credible* promises to withdraw. Would they, for example, set stiff penalties for themselves? The LDCs are not, typically, able to monitor the communication systems of the developed countries, so who will act as policeman?

Leasing Unused Orbit Slots and Spectrum

"At the other extreme, just short of a detailed *a priori* orbit spectrum plan, stands the possibility of permitting the holders of unused rights to lease them to other nations. . . . This option permits the LDC leaders to bring home symbolic success in wresting a 'rightful' share of orbit spectrum rights with allegedly great economic value to others and to lease them to the highest bidders."

However, Levin questions the value of those slots and spectra. "The power, orbit location, frequency, bandwidth, antenna beam, directivity, etc., each nation was authorized to use at the ITU Broadcast Satellite Conference in 1977 were specified so narrowly that there were few alternative uses or users for which they would be suitable."

Periodic Planning Conferences

This idea is for conferences every four or five years that draft eight-year plans. Initial allocations would be limited to immediate needs and expected lifetimes of existing satellites; future allocations would be postponed until the technologies develop.

Levin worries about "grandfathered" rights (those previously allocated and carried over) interfering with new conference agendas. He calls bilateral trades under these circumstances "awkward and few at best," and notes that they would be effective mainly between "long, narrow countries with extensive common borders," such as Argentina and Chile.

Regional Consortia

Users of common pool resources may choose to have their government enter into regional consortia "to restrain the proliferation of domestic satellite systems and hence the race for slots and frequencies. We aready know, for example, that the 104 members of INTELSAT would individually claim far more orbit spectrum for domestic systems than they collectively would if INTELSAT opted to provide the domestic service itself. . . . The big question is what the LDCs would get in return for their cooperation in economizing orbit slots or spectrum."

Levin gives three options for regional consortia. One is a system of exchanges of allocated orbits and spectra among the participants—which could run into the same limitations as the other schemes for resource-sharing. Another is for an LDC to "hire an outside concessionaire on a lowest bid basis . . . without exclusive rights [to the resources]." The third alternative would be for an independent "concessionaire" to operate the space segment and to take on more than one LDC as its clients. However, Levin is not convinced that LDCs will be warm to that idea, since they would lose sovereignty. Indeed, as he says, "the likelihood of any such consortium materializing may be impaired by the internal divisions and rivalries of members within the region."

An INTELSAT Subdivision

If INTELSAT provided domestic service for developing nations, orbit resources would be conserved. But would that "induce LDC members to abandon claims for future systems? Or would LDCs reject continued dependency on INTELSAT in view of its traditional focus on high-volume traffic in major urban corridors, with heretofore low-power satellites and high-cost earth terminals? Perhaps only if LDCs ultimately perceive monetary returns on their INTELSAT investment (as co-owners/users) might their own direct claims on orbit slots and spectrum be foregone."

Ultimately, Levin doubts that less-developed and highly developed nations will soon agree to implement any of these options. He hopes instead that his "five middle range options offer starting points for identifying workable compromises between the current evolutionary management system of first-come first-served which Third World nations deplore, and the kind of minutely specified *a priori* planning they increasingly appear likely to impose."

Will It Work?

INTELSAT's Joseph Pelton, assistant to the director-general, believes that cooperation and compromises—not *a priori* planning—will provide the solutions. As technology improves, more and cheaper communication channels have become available, making it somewhat easier for developing countries to take advantage of them. "Essentially, the world has learned to depend on telecommunication systems," he says. "Most countries feel comfortable leasing services from INTELSAT, or Intersputnik, and that kind of cooperation is a good way to proceed. I don't believe that economic measures will work as well."

Levin, an economist, responds by saying that INTELSAT offers only a "technological 'bailout' from the hard choices"; but he also admits that no amount of planning nor any allocation scheme will change the fact that some orbit slots are already full.

Heather Hudson helped small countries set up satellite links while she worked for the U.S. Agency for International Development. She says that "we shouldn't rush headlong into the American-style 'market' approaches, because the rest of the world feels more comfortable with an 'administrative' system." (That is reminiscent of Dr. Ploman's common-law/civil-law dichotomy.) She goes on to say that the orbit arc and spectrum bands are "social resources, tied to the role of telecommunications in a society and to that society's development. Carriers want flexibility to respond to new demands; users want affordability, and lower costs.

"The North American model of competition only works where there are alternatives, and yet some prior allocations have proven to be useful in the U.S., such as those setting aside portions of the FM radio and UHF television bands for educational organizations years in advance of their abilities to use them, and even before most such groups had defined their needs. Now," she says, "the spectrum is there to accommodate new users as they come along."

In the contentions between political scientists and economists, there must always be room for these kinds of dialogs to emerge. The worst thing would be for the people involved to throw up their hands and say "it's impossible." As Hudson puts it, "We ignore arc and spectrum issues at our peril."

Chapter 6

PIRACY ON THE HIGH FRONTIER

One summer day in 1980, the Minister of Universities, Science, and Communications set up a satellite antenna in the backyard of his office at the Parliament Building in Victoria, British Columbia. The moment his receiver displayed American television programs, The Honorable Dr. Patrick L. McGeer had, in effect, broken Canadian law—if not its letter, at least its spirit.

"The law is an ass, if that's the law," he declared afterward. "If somebody wants to put a signal down in my backyard, nobody can tell me not to look at it."

The stunt drew attention to a question that will not be quickly answered: whose sky is it, anyway? The idea that the geostationary orbit is open to whomever wishes to park a satellite there—the policy the U.S. calls "open skies"—has divided the United Nations and regional treaty organizations; now it divides the U.S. from Canada, and aggravates an internal dispute between Canada's federal government and its provinces over who shall control the airwaves: the federal government—with its dedication to the country as a whole—or the provinces—as the closest government to the people?

Unlike the U.S., Canada has no constitution to resolve that question. The British North America Act of 1867 set up the present system which gives to the provinces jurisdiction over "Lines of Steam or other Ships, Railways, Canals, Telegraphs, and other Works or Undertakings connecting the Province with any other or others of the Provinces, or extending beyond the Limits of the Province."

Like the U.S. Communications Act of 1934, Canada's Radio Act of 1935 was written before new technology burst out of those laws' constraints. The federal government in both countries took on the job of regulating telecommunications, but the BNA Act gives the provinces more power vis-a-vis the federal government than the American states have, and in many ways they have begun to exert it. Canadians are now drafting their own constitution; domestic communications is among the most hotly debated issues, right up there with bilingualism, and rights to natural resources.

THE FIRST "WIRED NATION"

The problem really begins in viewers' homes. Since most Canadians live within 100 miles of the U.S. border, it has always been easy

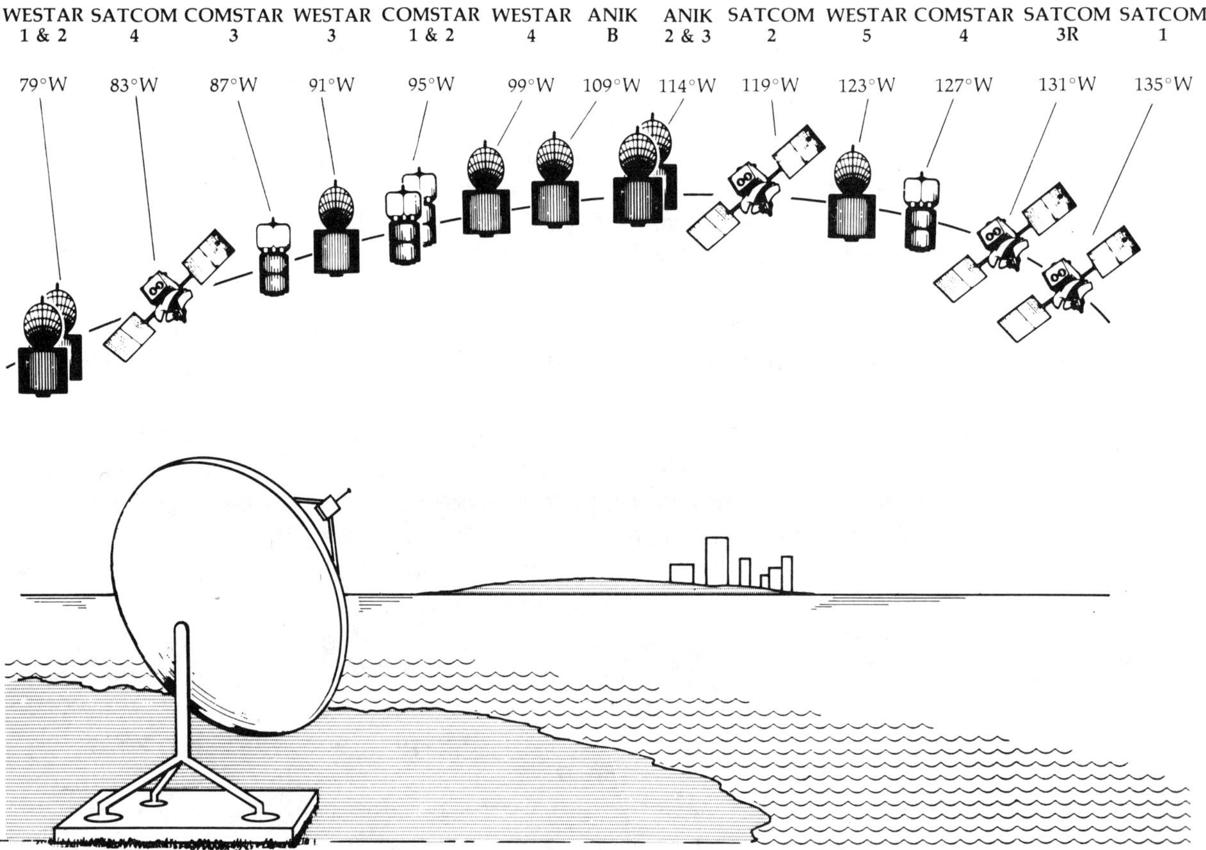

Broadcast satellites and

for them to watch American broadcasts. To encourage cultural diversity, and especially to promote Canadian programming, Canada was the first major nation to commit itself to cable tv, and the first to launch a domestic satellite system (Anik) for public education and entertainment. Now, virtually all of Canada's major cities and towns are heavily cabled. Urban Canadians get a wide offering of French and English channels and a lot of news and information programs both national and provincial. Yet, unlike their U.S. counterparts, there was no pay-tv until 1983, for movies, sports, etc. The reason is that along with the new technologies came a national requirement for a significant percentage of "Canadian content"—programming by and

				SATCOM 3									SATCOM 4		
EASTERN	CENTRAL	MOUNTAIN	PACIFIC	TR-4 SPLT Spotlight	TR-5 THE MOVIE CHANNEL	TR-10 SHOWTIME West	TR-12 SHOWTIME East	TR-16 HTN Home Theatre	TR-13 HOME BOX OFFICE West	TR-24 HOME BOX OFFICE East	TR-20 Cinemax East	TR-23 Cinemax West	TR-6 BRVO Bravo	TR-7 N.C.N ESCP Escapade	TR-8 TEC New York
6 30	5 30	4 30	3 30	Movie: 'Falling in Love Again' "	Prog Cont'd " Movie: 'Star Trek – the	Prog Cont'd " " "	Movie: 'Seems Like Old Times' "	Prog Cont'd " " "	Movie: 'Coast to Coast' " "	Moonchild " " "	Movie: 'Enemy of the People' "	Castaway " Irene Moves In			Sporting Chance " "
7 30	6 30	5 30	4 30	" " " "	Motion Picture' " "	" " Movie: 'Thunderbirds	" " " "	Travel Chan. " " "	" " " "	Beach Boys in Concert " "	" " " "	Blinker's Spy Spotter " "			Just For Fun " Animal Express
8 30	7 30	6 30	5 30	Movie: 'Tommy' " "	" " " "	in Outer Space' " "	Movie: 'Victory' " "	Movie: 'Don't Go Near the Water' "	The Gold Bug " "	Movie: 'The Other Side of the Mountain	Movie: 'Nightmare' " "	Little Detective " "	Performance: The Mikado " "	August Playboy Magazine "	Movie: 'Tomorrow' " "
9 30	8 30	7 30	6 30	" " " "	Movie: 'Sitting Ducks' "	Movie: 'Seems Like Old Times' "	" " " "	" " " Travel Chan.	Moonchild " " "	Part II' " " "	" " " "	Movie: 'Enemy of the People' "	" " " "	Movie: 'American Gigolo' "	" " " "
10 30	9 30	8 30	7 30	Movie: 'Smokey Bites the Dust'	" " Movie: 'Pictures'	" " " "	Movie: 'Seniors' " "	Movie: 'Clash of the Titans' "	Beach Boys in Concert " "	Movie: 'Victory' " "	Movie: 'Baltimore Bullet' "	" " " "	" " Movie: 'Pic- nic at	" " " "	A Family Affair " "
11 30	10 30	9 30	8 30	" " Movie: 'New Year's Evil'	" " " "	Movie: 'Victory' " "	" " Laura, Sweet Laura 1	" " " "	Movie: 'The Other Side of the Mountain	" " " "	" " " Movie: 'Back	Movie: 'Nightmare' " "	Hanging Rock' " "	August Playboy Magazine "	Movie: 'Tomorrow' " "
12 30	11 30	10 30	9 30	" " " "	Movie: 'Cheaper to Keep Her' "	" " " "	Movie: 'La Cage Aux Folles II' "	Movie: 'Don't Go Near the Water' "	Part II' " " "	Movie: 'The Night the Lights Went Out in	Roads' " " "	" " " "	" " Performance: The Mikado	Movie: 'American Gigolo' "	" " " "
1 30	12 30	11 30	10 30	" " Movie: 'It's My Turn'	" " Movie: 'The Pilot'	Movie: 'Seniors' " "	" " Movie: 'Challenge	" " " "	Movie: 'Victory' " "	Georgia' " " "	" Movie: 'Papillon' "	Movie: 'Baltimore Bullet' "	" " " "	" " " "	A Family Affair " "
2 30	1 30	12 30	11 30	" " " "	" " " "	" " Laura, Sweet Laura 1	the Dragon' " " "	Movie: 'Clash of the Titans' "	" " " "	Movie: 'Alien' " "	" " " "	" " " Movie: 'Back	" " " "	August Playboy Magazine "	Limited Edition: Limited Edition:
3 30	2 30	1 30	12 30	Movie: 'The Kids Are Alright' "	Movie: 'Zoot Suit' " "	Movie: 'La Cage Aux Folles II' "	Movie: 'Seems Like Old Times' "	" " " "	Movie: 'The Night the Lights Went Out in	" " " "	" " " "	Roads' " " "	Movie: 'Pic- nic at Hanging Rock'	Movie: 'American Gigolo' "	Kelly Montieth: Last of the Summer Wine
4 30	3 30	2 30	1 30	" " " Movie:	" " " "	" " Movie: "	" " " "	Sign Off	Georgia' " " "	Beach Boys in Concert " "	Movie: 'M' " "	" Movie: 'Papillon' "	" " " "	" " " "	Limited Edition Limited Edition

program channel guide.

for Canadians. What Canada feared from the U.S. is what Third World nations call "cultural imperialism."

"This is a matter of cultural life or death!" declares Charles Belanger, Executive Assistant to Francis Fox, Canada's Federal Minister of Communications. A bilingual Quebec citizen, Belanger says of the U.S.: "You're such a powerful nation, we Canadians will never be able to produce such an amount of programming. In news and public affairs we're doing well, but in drama and entertainment we're overwhelmed by the American presence."

The Canadian Radio-Television and Telecommunications Commission (CRTC) can regulate cable tv, and can force licensed broadcast stations to air Canadian content, but it has long been aware that in rural areas, far from the U.S. border and beyond the reach of most broadcast tv signals, satellite dishes have been used to receive American pay-tv movies and "superstations." RCA's Satcom I serves Alaska, so its footprint necessarily reaches into western Canada. In many small towns there, enterprising people have bought receiving antennas; where few broadcast tv signals are available, many towns have installed cable tv systems or low-power tv transmitters to distribute the signals. Like Dr. McGeer's dish behind the Parliament building, theirs too are illegal.

TV IN THE FRONTIER NORTH

Fort St. John is typical of those northern British Columbia towns: an oil and forest products community which receives only two weak broadcast stations. Station CJDC comes from nearby Dawson Creek, but "it's a pretty poor channel," says Dennis Jacobsen. "The colors are off, and their video tape recorders are antique." CFRN is "bounced 700 miles" through repeaters from Edmonton, Alberta. Jacobsen is a civil engineering technologist who admits, "Before I started, I knew very little about satellites. Now I can see it's very easy—like a large stereo set—you just have to use a little more care to make sure your connections are good." In only a few hours, he and the other residents of Fort St. John set up a turnkey system they'd bought from a supplier in the town of Kamloops.

"It was a community decision," he explains. "We asked for voluntary membership fees to buy the antennas: $45.00 for a family and $20.00 for a single person. We collected $95,000." They bought two dishes and four 10-watt (low-power) VHF transmitters to distribute the satellite signals. Local viewers, by acclamation, decided to watch

BCTV from Vancouver via Anik, and The Movie Channel, Cinemax and WGN (Chicago) via Satcom I.

Jacobsen denies that this is "piracy," because the satellite's signals have crossed an international boundary. The American FCC requires cable tv, master-antenna tv and other distributing systems' operators to pay a monthly fee, per subscriber, to the program provider. But since the movie and "superstation" channels are not licensed by the CRTC to operate in Canada, they cannot do business there. Says Jacobsen, who has tried to pay for the services: "They won't accept our money." In fact, Home Box Office and other pay-tv companies have consistently refused to accept money from *all* independent, would-be subscribers. They prefer to deal with CATV or MATV systems, which offer them hundreds or thousands of customers in a single transaction. While it might be an accountant's nightmare to cope with many small customers, it is surely an attorney's nightmare to decide who is a legitimate customer and who is not.

Canada's federal government, in an effort to slow the growth of American television reception, has authorized a pay-tv service of its own. Dubbed the "Canadian Package," CANCOM costs each subscriber $4.00 per month to get BCTV (Vancouver), CITV (Edmonton), CHCH (Hamilton, Ontario), and TCTV, the French station from Montreal, via Anik. Jacobsen is skeptical. "They expect all the illegal dishes to swing around and apply for licenses to become legal. We're not pleased. The Canadian channels will each cost a dollar a month, while the fee for WOR (New York) or WTBS (Atlanta) is only about ten cents a month. They want us to watch Canadian content and pay ten times as much for it! What *we* want is for the American channels to be legalized."

He speaks not only for the people of Fort St. John, but for the 75 other communities in B.C., Alberta, Saskatchewan, the Yukon, and the Northwest Territories that have organized themselves into the Western Canada Satellite TV Association, of which he is president. In one newsletter, the Association complained that the stations participating in the Canadian pay-tv offering "will not contribute towards the rental of the satellite time from their advertising revenue," as the American "superstations" do; so Western Canadians will have to foot the bill. The Association is polling all its members to see if they want to boycott the Canadian package; Jacobsen thinks they will.

Transborder satellite traffic is not impossible. In 1982, the two countries agreed to share a satellite network for business communication. Earth stations owned by American Satellite Co. and Telesat

Canada, respectively, would be able to share computer data and other digital services, including digitized voice traffic. The customers for the system would be corporations whose business was in both countries, such as mineral extraction and energy production. Commercial television programs, however, are specifically excluded from the agreement.

THE "RADIO POLICE"

Francis Fox, the federal Communications Minister, has said that private tv viewers shouldn't need a license or other government permission to watch satellite programming, and that the policy toward rural areas with illegal dishes will be one of leniency. But, as Charles Belanger elaborates, "You'll never have this government advocating an 'open skies' policy. We're not contemplating Army action, but if cable or broadcast providers deliver more than their licenses permit, they will get into trouble."

In October, 1980, the Royal Canadian Mounted Police seized the components of a satellite receiver from an apartment house in Burnaby, a suburb of Vancouver. The owner, who had provided satellite signals as well as cable service to his tenants, was charged under the Broadcasting Act (part of the Radio Act) for operating a radio apparatus without a license. Unfortunately for both the Crown [prosecutor] and the defense, no precedent was set by the outcome of the trial.

What happened was that in May, 1981, a Provincial judge dismissed the charges over an issue neither of them had raised, and without a motion from the defense. The Radio Act protects from interception signals defined as those "propagated in space without artificial guide." That is, it covers only broadcast signals—as distinguished from signals sent through cables, wires, etc.—in the traditional interpretation of the law. Witnesses representing the satellite industry had made many references to "directional" antennas and other guiding mechanisms, in an effort to demonstrate that the signals were proprietary, private, and not intended for "public" reception. The judge, however, took that to mean that satellite signals *were* artifically guided, and so did not fall under the protection of the Act.

P. W. Halprin, Special Prosecutor for the Department of Justice, said in justifying his appeal to a higher court, "The defense wanted the issue decided as much as the Crown did; I think they were as much surprised by the judge's ruling as we were." If the judge is upheld, he admits, "it will take a Parliamentary amendment to the Act" to change it. Halprin, who specializes in criminal law, warns that, "In the meantime, anyone in Canada operating such equipment is in jeopardy of having it seized."

Defense counsel Barry Adams playfully calls the Department of Communications' enforcement officers "the radio police," but he seriously criticizes their selective enforcement policy of raiding a southern Canadian installation and not a northern one. "Licensees are easy to regulate," he says. "In Vancouver, which is 90 percent cabled [and hence, under federal license], the Department of Communications can control the new forms of technology. But there are a lot of dishes out in the country, in little towns, and a lot of Members of Parliament whose constituents have dishes. Besides, imagine a 142-pound enforcement officer going into a northern B.C. mining community to take action!"

Robert Stewart, Deputy Minister under Patrick McGeer, says that a member of Parliament actually got a call from a rural town mayor saying he—the mayor—couldn't be responsible for the safety of any Department of Communications inspector. McGeer chuckles at that, insisting that "The word 'illegal' depends on which way you point the dish. Toward Anik is all right, but when we pointed our dish at the American satellite, the Canadian government said we were stealing signals. So we called some of the programmers. WTBS [Ted Turner's Atlanta 'superstation'] said it was okay; Southern Satellite Systems—WTBS's common carrier broker—said they were happy to see Canadians watching. The Christian Broadcasting Network said 'hallelujah!"

He advises the federal government: "If a signal is falling on your property and you don't want it, ask them [the Americans] to take it off. If you want to compete, then turn out better programs. When the British make a good movie it's seen around the world. When we turn out good television, it'll be watched in the U.S."

In Ottawa, Belanger insists that the federal government "is not contemplating policing to get our people to be attracted to our own stuff. We're going to stress incentives." All-Canadian productions, he says, would be encouraged by more grants to the arts, "fiscal methods" including changes in the tax laws, and greater cooperation with the private entertainment industry. The National Film Board of Canada is a good model for cooperation, he admits. Many of its films have been successful worldwide and regularly win critical acclaim.

"We're trying to show openness, but we want to get the best for our citizens, to enhance our own people. In southern Canada we're spoiled by watching American programming."

McGeer, in Victoria, fears the federal government will try to "turn the technological clock backward to protect incompetency. There's no point in setting up a philosophy like 'Canadian content' that technology makes impossible. There's no serious attempt to enforce the law right now; even those low-power VHF repeaters are openly winked at. Don't get me wrong—Francis Fox and the CRTC people are nice fellows, but they're committed to regulation and I'm not."

"The future," McGeer says, "will bring us hundreds of TV stations over cable and direct-broadcast satellites. Specialty networks will cover sports, education, and hobbies. Local news will come back—after all, newspapers didn't die when they invented the Associated Press. The ones who are *really* hurting from satellite technology are the American TV networks. They'll be obsolete."

WHAT'S SO DIFFERENT ABOUT CANADA?

To U.S. citizens, Canadian controversies may seem bizarre; after all, Canada is not so very different from the U.S., is it? On both sides of the border people look and talk alike, drive the same cars, eat the same foods. . . . Canadians are our northern cousins, as the schoolbooks tell us, right?

Not exactly. Mike Israel is an officer of Teleglobe Canada, the country's international telephone carrier, and he reminds people in the U.S. that Canada has a completely different terrain; its two major industrial clusters are a continent apart; and the vast majority of its people are concentrated in a narrow strip of land 3000 miles long. "We may both look alike, but where telecommunication is concerned, we are much less alike," he says. "As in developing countries, our emphasis has been to provide for the basic needs of people in remote areas and—amazingly—that opened us up. It actually encouraged the authorities to develop new, communication-based health care and educational services for *all* our people.

"Our population is only one-tenth that of the U.S., so we cannot regard policy or regulatory issues in the same light as the U.S. does; we don't have the revenue base. Canada's telecommunication infrastructure is a mosaic. Some phone companies are owned and operated by the provinces; others are owned and operated privately; but they're all federally regulated. Because of our cultural diversity, some provinces feel they have to regulate television and radio content. 'Can-con' [Canadian content] is supposed to apply to all the provinces but, philosophically, it's a way to promote local industries and talent in those provinces."

The issues have come to a head in Canada mainly because its people are more aware, on a day-to-day basis, of the telecommunications links within their country and those to the rest of the world. Mike Israel acknowledges that, like the U.S., Canada is and wants to continue to be a leader in communications technology, and he foresees the possibility that the U.S. may not be so isolated from other countries in the future, over the "open skies" issue, as it is now. "Things may change," he admits.

But to the people in the British Columbia countryside, in towns like Port Hardy and Port MacNeil on Vancouver Island, and in the communities of the Peace River—Hudson Hope, Chetwynd, Dawson Creek and Fort St. John—for those people, the debate over "open skies" is moot. They're not talking about it. They're living it.

Section 3

POLITICS AND SOCIETIES

"Nation shall speak unto nation"

—motto of the British Broadcasting Corporation, 1927

Chapter 7

INTELSAT

Less than 20 years after its founding, the International Telecommunications Satellite Organization (INTELSAT) is at once the most powerful and the most vulnerable communications carrier in the world. Two out of every three international telephone calls now pass through its satellites. Yet on every continent, as local economies expand, INTELSAT's monopoly is being eroded by domestic and regional satellite systems. As AT&T finally learned: a giant, unified communications organization—however efficient—may be out of step with technology, economics, and the aspirations of its customers.

THE AMERICANS' GIFT

The consortium was formed in 1965, in response to a U.N. resolution a year earlier calling for satellite-borne communications to be "available to the nations of the world as soon as practicable on a global and non-discriminatory basis." The U.S. delegation wanted to call it the "Global Telecommunications Satellite Consortium," but Motoichi Masuda, a Japanese delegate, complained that the acronym "GLOTELSAT" sounded like the word "gurotesuku," meaning "grotesque." The delegates adopted "INTELSAT" instead.

Adolfo Alessandrini, who headed the Italian delegation at INTELSAT's founding, credits "the deeply constitutional and liberal American mentality," for the success of INTELSAT.

Jose Alegrett, now director of INTELSAT's external relations, says, "In a way, I believe INTELSAT was a gift of the United States to the world. At that time, though, I felt that somehow the United States or COMSAT had misled my country by convincing us that Venezuela should become a member of this organization, even though it would be years before we could become a user of the system."

As a delegate to the 1969 Plenipotentiary Conference, Alegrett was worried about the controversy over INTELSAT's governance. Two alternatives were suggested by the preparatory committees: "a commercially pragmatic approach, with the management of the system being contracted out to COMSAT; and a more political view, with much of the power in INTELSAT being vested in an equivalent of the present Assembly of Parties, and with a permanent international management organ of its own." The eventual compromise was drafted by an unexpected cooperation between Australia, siding with the commercial approach, and Japan, allied with the opposition; after a bit of tinkering, it was accepted by the others, and INTELSAT had a truly international character.

THE SIGNATORIES

INTELSAT is founded on two documents: the *Agreement* which makes a country a "party" to the Agreement, and the *Operating Agreement* which must be signed either by a country's government or by its designated telecommunications entity; whichever signs becomes the "Signatory."

The U.S. created a quasi-governmental organization called the Communications Satellite Corporation (COMSAT) to be its signatory. In some ways, it was similar to the "authorities" that build bridges and tunnels in major cities. Like them, it enjoyed exclusive rights to operate a facility (satellite earth stations), to collect tolls (tariff rates) from users, and to be both a privately owned corporation and a government-sanctioned monopoly at the same time.

COMSAT is "publicly held" in the American sense: it is owned by stockholders. Other signatories, however, are agencies of their governments or, in many cases, are governments themselves. Japan, for example, is represented by Kokusai Denshin Denwa Co., Ltd. (KDD), and Chile by Empressa Nacional de Telecommunications S.A. (ENTEL), both of which are post-and-telegraph monopolies. France, Austria, Mexico, Chad, Egypt, Israel, and many others represent themselves directly through government bureaus.

INTELSAT sells its capacity—that is, time on its transponders—for a fee that covers operating expenses and depreciation, plus a markup; then it disburses a 14 percent return on investment to the signatories, proportional to the size of their shares.

The U.S. is the largest shareholder, with 24 percent. The U.K. holds 13 percent; France has nearly 6 percent; Germany, Australia, Japan, Brazil, and Saudi Arabia each have more than 3 percent; Italy and Canada each hold over 2 percent; and Iran, Argentina, Nigeria, Singapore, South Africa, the United Arab Emirates (UAE), Venezuela, Switzerland, and the Netherlands each hold something over 1 percent. The rest have less than 1 percent apiece.

Politically, the Organization of Economic Cooperation and Development—representing "developed" countries—controls two-thirds of INTELSAT, though many of its members hold shares smaller than 1 percent. Of the non-OECD countries, which represent the "other" third, only Brazil, Iran, Nigeria, Saudi Arabia, Singapore, the UAE, and Venezuela, hold 1 percent or more. Although China and Yugoslavia are members, the U.S.S.R. and its allies are not represented in INTELSAT; they have their own satellite system called Intersputnik (explained later).

Since use of the system was unequal (the industrial countries used INTELSAT far more than the less-developed countries did), it did not seem fair to divide either expenses or profits equally among the signatories. By 1973, the consortium adopted a system of "investment shares," according to Reginald Westlake and Carl Reber, financial officers of INTELSAT and COMSAT, respectively. These shares reflect each country's percent of use, and are recomputed annually; on occasion, millions of dollars change hands as some countries, in effect, "buy" shares from other countries. As its financial base has broadened, accounting changes keep members' investments consistent with their use of the system.

INTELSAT circuits are used regularly for telephony, telegraphy, computer data, and television, and occasionally for teleconferencing, emergency broadcasting, and distribution of radio programs. The structure of INTELSAT tariffs permits a service to be preempted by traffic with a higher priority, if the need arises; higher priority services pay higher rates. All charges are based on the "unit of utilization" which is the equivalent of one-half of a telephone circuit (i.e., one "end" of a two-way conversation), occupying about 4 kHz in bandwidth. Multiples of that "unit" are calculated for billing more sophisticated services.

THE GLOBAL VILLAGE

The growth of INTELSAT has been dramatic. In 1969—the first year in which the Atlantic, Pacific, and Indian oceans were all served—there were about 2500 half-circuits ("units") available. By 1981, more than 50,000 were in use. Since 1972, INTELSAT services have been reliably available for use—that is, without technical interruption—more than 99 percent of the time.

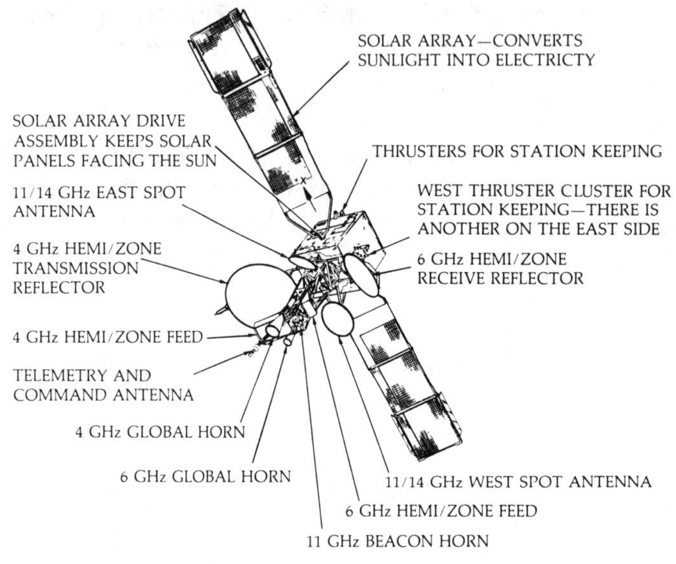

INTELSAT V Satellite (simplified)—Courtesy NASA

Joseph Pelton, executive assistant to the director, gives this account of INTELSAT's growth. The INTELSAT IV system, in 1975, could carry 6000 telephone calls, plus two color tv channels. INTELSAT V, which was launched in 1981, carried 12,000 telephone calls, plus two tv channels. In 1986, INTELSAT VI satellites will be able to carry 36,000 telephone calls, plus two tv channels. If INTELSAT VI were to be used exclusively for digital computer data, it would be capable of transmitting some three billion bits of information every second. "This is sufficient capacity, should the need arise, to transmit the Encyclopedia Britannica some twenty times a minute," says Pelton.

INTELSAT's earth station antennas have been made more efficient, too. The first standardized size (called Standard A) was 30 meters in diameter and—though very efficient—they required a large volume of traffic to pay for their cost. They are still used by developed countries, such as the U.S. and the U.K. Smaller countries use smaller earth stations: many are 11 meters in diameter (Standard B). Now, newer and smaller dishes (Standard E) are as small as 3.5 meters. Obviously, there is a trade-off. When the dish is smaller, it

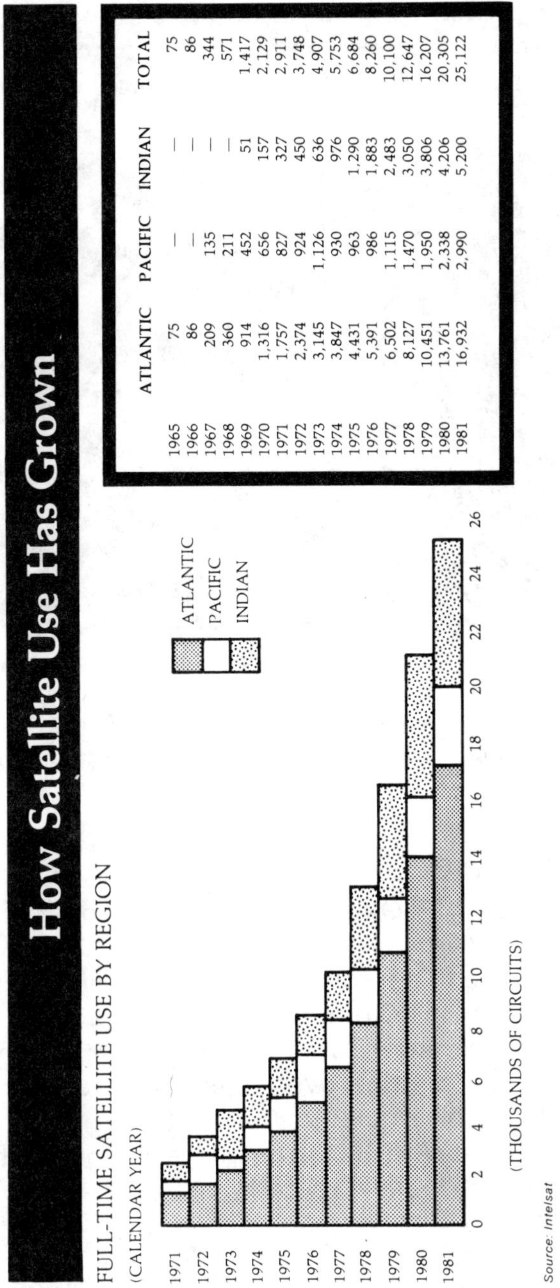

	ATLANTIC	PACIFIC	INDIAN	TOTAL
1965	75	—	—	75
1966	86	—	—	86
1967	209	135	—	344
1968	360	211	—	571
1969	914	452	51	1,417
1970	1,316	656	157	2,129
1971	1,757	827	327	2,911
1972	2,374	924	450	3,748
1973	3,145	1,126	636	4,907
1974	3,847	930	976	5,753
1975	4,431	963	1,290	6,684
1976	5,391	986	1,883	8,260
1977	6,502	1,115	2,483	10,100
1978	8,127	1,470	3,050	12,647
1979	10,451	1,950	3,806	16,207
1980	13,761	2,338	4,206	20,305
1981	16,932	2,990	5,200	25,122

The number of INTELSAT circuits rose from 75 to more than 20,000 in its first 15 years. Most customers want transatlantic service, but demand in the Pacific and Indian Oceans is steadily increasing.

requires more power from the satellite's transponder. INTELSAT charges 50 percent more per half-circuit if a country has the Standard B than if it has the Standard A; comparable rate adjustments are made for smaller dishes.

***Standard A:** The original COMSAT/INTELSAT earth station at Andover, Maine, is still one of the largest in the world. The size of the dishes is known as "Standard A" and is required for high-volume traffic routes, such as transatlantic service. The dome in the background provides another dish with protection from rain and snow. (Courtesy COMSAT)*

TECHNOLOGY IN THE DRIVER'S SEAT

The dominance of COMSAT, among the signatories, has given INTELSAT some important technological "firsts." COMSAT was the official carrier for the first transatlantic messages on Early Bird. Holding the largest share of INTELSAT, it has been able to guide the design of new satellites and related technologies, helping to give the U.S. the technological edge over most other countries in space. When an INTELSAT bird failed to reach its orbit slot, an American experimental satellite (ATS-3) was made available to broadcast the Olympics from Mexico City; when INTELSAT's Atlantic satellite F-2 stalled, the old Early Bird was dragged out of retirement in 74 minutes; in less than four hours, users could not tell that anything had changed. The biggest tv audience INTELSAT ever assembled was the one that watched the Americans land on the moon—500 million people. At least it was the largest until the 1982 World Cup soccer match.

Although the U.S. and some other developed countries have their own satellites for domestic communication, most countries cannot afford to design, build, launch, and insure one; few of them have the local telephone networks in place to keep the transponders busy. Algeria was the first country to use INTELSAT circuits for domestic communication; that was in 1975. As with many developing countries, its population clusters are separated by virtually uninhabited regions, the terrain is unsuitable for microwave towers, and the cost of stringing cables is prohibitive. A network of earth stations, however, focused on a common satellite, is cost-effective. At first, INTELSAT leased only "spare" time and transponders for domestic traffic, and reserved the right to preempt those transponders for international customers. Now, since so many countries want these "instant" networks, INTELSAT has changed its technical requirements for future satellites: domestic users are included in the planning process.

SPUTNIK

The Russian word for "satellite" entered the vocabulary of the rest of the world October 4, 1957, and forced the U.S. to embark on a crash-course in science and technology. The "space-race" is rather a draw, now, with the Americans mapping the Moon and Mars, and the Soviets doing the same for Venus. The intent of both sides is clear: military advantages on earth. Long-term survival in orbiting laboratories seems to be the Russians' goal; piloting reusable shuttles is the Americans' forte. Once, the two countries linked up in space, and televised the event all over the earth, but nowadays the rivalry is too intense for that.

The word "sputnik" survives, though, in "Intersputnik," the Soviets' answer to INTELSAT. Their country is the largest in the world, extending across six time zones, and their geopolitical concerns are global; so it is not surprising that they have built a system more or less equivalent to INTELSAT, with some special features to suit their geographical anomalies.

According to Theo Pirard, European correspondant for *Satellite Week* newsletter and *Satellite Communications* magazine, there are 14 signatory members of Intersputnik: the Soviet Union, the six Warsaw Pact nations, Mongolia, Vietnam, South Yemen, Afghanistan, Syria, Laos, and Cuba. Algeria, Iraq, and some other countries use the network, though they are not signatories, to communicate with the Soviet Union and its allies.

Because the U.S.S.R. extends beyond the Arctic Circle, and its major cities are much further north than those of Europe or North America, the first Intersputnik network used satellites in low-earth, polar orbits which had to be tracked continuously. But that Molniya series was replaced, in the 1970s, by the geostationary Gorizont series, which uses the C-band (6/4 GHz). North American back-yard satellite dishes that are east of the Mississippi river can pick up the Atlantic satellite Gorizont Statsionar 4, at 14° W, which transmits television programs to and from Cuba.

Countries in the Intersputnik system pay much less for transponder time and services than INTELSAT customers do. According to Pirard's reports, an Intersputnik voice circuit can be leased for $11,615 a year (1982 U.S dollar equivalent), while INTELSAT charges $19,358. Earth station charges are comparably scaled. The low prices, however, do not make up for Intersputnik's limited range and lack of terminals in the rest of the world.

THE SUPERPOWERS' POLE STAR

An unusual feature of Intersputnik, by the way, is that it includes four Molniya polar satellites in a highly unusual orbit to serve the northernmost reaches of the U.S.S.R. According to Mark Long and Jeffrey Keating, authors of *The World of Satellite TV,* the Molniya orbit is shaped like a boomerang, with the satellites following a bent figure-eight track that it takes 736 minutes to cover. When they are closest to the earth, near the South Pole, they are only 467 km (291 mi) up; when they are at the apogee, over Hudson's Bay, they are 40,813 km (25,508 mi) high. Like baseballs thrown straight up, they seem to "hang" at the highest place. For about six hours, a relatively simple tracking mechanism can be used to keep whichever Molniya is then at apogee in sight; Long and Keating give satellite-hobbyists directions for tracking them.

But the Molniya system has a strategic importance: it is part of the U.S.–U.S.S.R. "hotline." Until 1976, the hotline was carried exclusively by ITT cables from Washington to New York, under the Atlantic to London, under the North Sea to Copenhagen, Stockholm and Helsinki, and then overland to Moscow. The cable was subject to accidents, including a manhole fire in Baltimore, an Atlantic fisherman's net, and a Finnish farmer's errant plow. Back-up radio was available, but the superpowers decided to use satellites instead.

The Soviets keep a channel open for the hotline on whichever

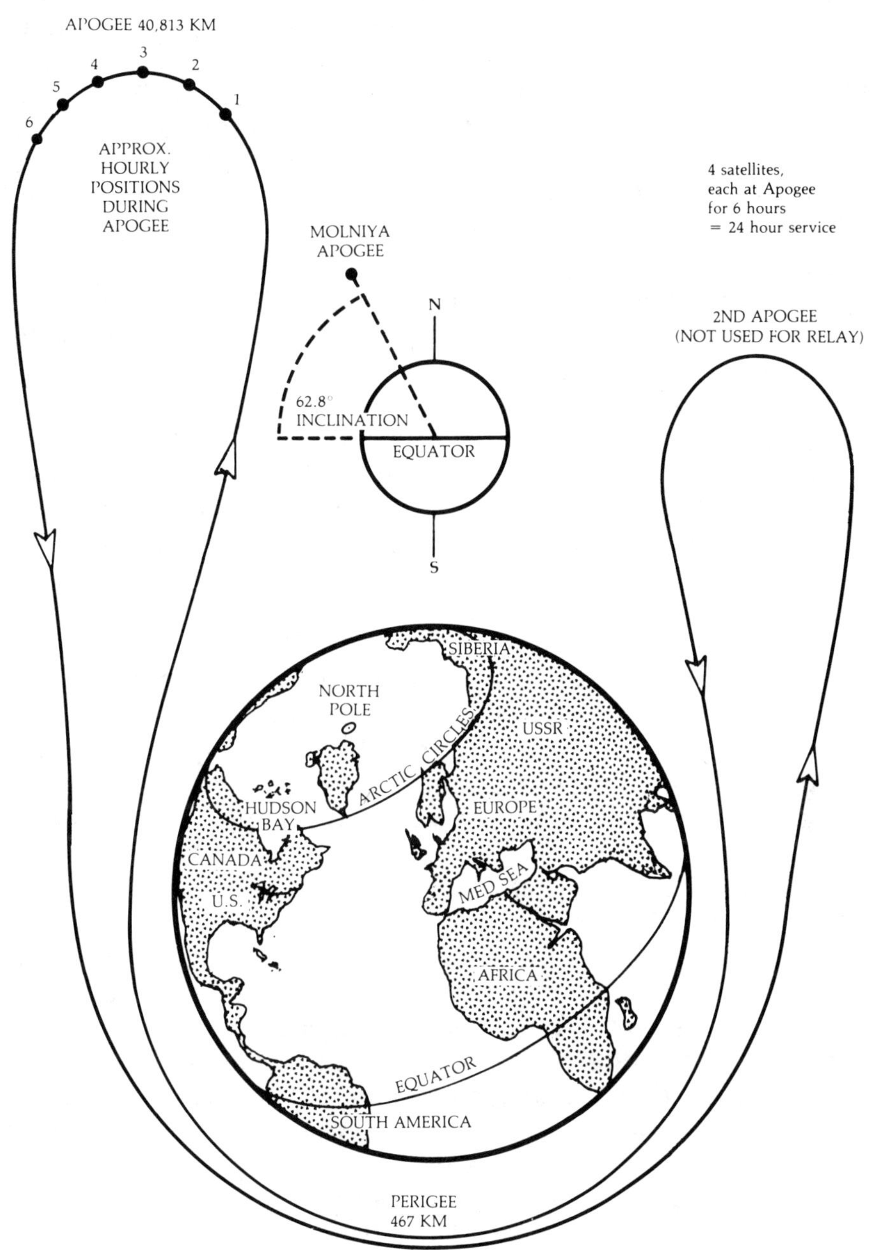

Orbit of the U.S.S.R.'s Molniya satellites.

Molniya is over North America at that time. From an earth station in Ft. Detrick, Maryland, the Americans track the same Molniya and keep that channel available for hotline traffic. Redundancy is built into the system by also holding open a leased INTELSAT circuit on the bird over the Atlantic, which can send messages via Europe. Both circuits, like their cabled predecessor, use printing teletypewriters—not voices—so that there will be no confusion about the nature of a message. The Molniya circuit operates in Russian, while the INTELSAT circuit uses English; translators are on duty on both ends 24 hours a day.

Pirard says that Intersputnik is experimenting with 14/11 GHz K-band satellites, and smaller earth stations that—like those in INTELSAT's plans—will allow less-developed countries to lease or purchase service more easily. "In the near future," Pirard says, "Intersputnik may well prove to be a strong competitor for the world market." Other people are not so sure, and one reason may be the rise of regional and domestic satellite networks that will drain away some of the business from both INTELSAT *and* Intersputnik.

LOCAL COMPETITION

INTELSAT may soon lose some of the business it helped to create, to the very countries who formed INTELSAT in the first place. Developing countries are building their own systems and—in partnerships with adjacent countries—establishing international networks of their own under the rubric of "regional" satellite consortia.

Consider Indonesia. That country has had its own satellite, called PALAPA, since 1976. It could justify the cost of building a domestic telecommunication network only if it could reach the remote places where people worked to extract oil and minerals; that way, the largest user fees would be borne by transnational mineral and oil companies—the people best able to pay for them—and rural villages could get educational tv and some public telephones more or less for free. Military communications were also built into PALAPA from the very beginning. PALAPA was so successful that a second satellite was launched in 1983. With the added capacity, Indonesia will be able to sell telecommunication services to its Southeast Asian neighbors.

Domestic and regional systems are proliferating in many places. Australia's domestic satellite system is expected to have a spot-beam aimed at Papua, New Guinea. The Moslem states of the Middle East have an ARABSAT project underway, and there is talk of an African satellite as well.

If that were all, INTELSAT would probably not mind; those are fairly thin-route areas. But if Europe had a regional satellite, INTELSAT could suffer a considerable loss of revenue. So it is watching the development of EUTELSAT with great interest. EUTELSAT is a provisional organization whose members are post-and-telegraph administrations; it was formed in 1982 with the goal of operating a satellite network for European telecommunication in the 1980s. EUTELSAT is expected to carry voice, telex, data, and television, but signatories expect to get most of their business from business users.

Europe, as a continent, has suffered through wars over territoriality, but is also produced the ITU and has kept it going through war and peace for nearly 120 years. Europeans have disagreed over many things, and satellites will not salve their wounds. Some European governments are upset at the prospect of receiving their neighbors' television programs. Conflicting codes for product advertisements, for example, or varying tolerances for nudity and other behavior, make the success of European direct-broadcast satellite television somewhat shaky. Belgium, Sweden, and Denmark ban tv commercials completely; France will not let children endorse products, and Austria won't allow children to appear in commercials, period.

Even if EUTELSAT concentrates only on business communication, it will have to resolve serious conflicts over "trans-border data flow": the legal restrictions on computer data exchanged among countries. It's a can of worms. For example, if a German national unscrupulously taps into a French computer operated by an American company, and uses what he finds to fleece the account of an Italian residing in Switzerland, who has jurisdiction to prosecute him?

Aside from that—and that's a book-length topic in itself—there are other sources of friction throughout the continent. Sweden, for example, imposes tight export controls over data that is even remotely "personal," such as credit history, while the U.K. has practically no laws on the subject at all. Countries with subsidized domestic communication industries, such as the one in France, will not easily tolerate competitive American or Japanese equipment sales. Luxumberg and Belgium—small countries reluctant to be overrun—want to limit the entry of U.S.-based communications carriers to Europe, in light of the American FCC's lifting of practically all restrictions. The Scandinavian countries actually told the American carriers in 1982 that, if they were to come to Europe, they would have to *bid competitively* against local carriers to provide new telecommunication services.

And EUTELSAT has an international problem all its own. There is some reason to believe that, by creating EUTELSAT, signatories would be violating one of their INTELSAT agreements. Article XIV of the *Agreement* permits Parties [countries] or Signatories to set up or join "space segment facilities separate from the INTELSAT . . . facilities." They are required to notify the INTELSAT Board of Governors so that technical and radio-frequency compatibility is ensured, but there is also this requirement: "to avoid significant economic harm to the global system of INTELSAT."

Andrea Caruso, secretary-general of the provisional EUTELSAT organization, was formerly an INTELSAT executive, and assisted in the drafting of its agreements. Now he asserts that new technology has made Article XIV seem archaic, if not irrelevant. In a presentation to satellite carriers in Washington, D.C., he threw down this gauntlet:

> Situations evolve relatively slowly in the field of politics or history, but the same cannot be said of the technological sphere and its applications. In the case of international conventions on the operation of advanced technology, would it not be desirable, even wise, to redefine at regular intervals and also to explain the terms used in the adopted texts? The Signatories could then agree on the interpretation of the treaty without questioning the substance of the treaty itself.
>
> INTELSAT was born of the willing cooperation and economic realism of 106 countries. What credibility can the Organization have . . . if its members are fettered by articles which were drawn up in an economic and technological context bearing no resemblance to the present situation and, moreover, which are interpreted with quasi-Draconian rigidity? Let us hope, for the general good, that the spirit of Article XIV will triumph over the letter, thus preventing the Article XIV syndrome from developing into an incurable disease.

INTELSAT officials express hope that EUTELSAT can fit, somehow, into the international framework of INTELSAT without causing an irreparable rift, just as INMARSAT has done. But the tension is there. Caruso's wish that "the spirit" of the law should prevail over "the letter" reflects a larger, philosophical difference of opinion from that of INTELSAT. The ITU itself is divided between developing nations which want "the letter" of an *a priori* spectrum and orbit slot allocation to prevail over "the spirit" of an *a posteriori* allocation claimed by the industrial nations.

In Chapter 5, conflicts in formulating Space Law were shown to be rooted, in part, on whether a country is predisposed toward "common law" or "civil law": a dichotomy comparable to "spirit" and "letter," *a priori* and *a posteriori.* Fundamental, philosophical disparities such as those do not allow for easy agreements, nor do they speed resolution of thorny international disputes.

Chapter 8

INMARSAT

The 1940s radio commentator Walter Winchell began his broadcasts with the salutation, "Good evening Mr. and Mrs. North America, and all the ships at sea." That was no exaggeration. Cables and local radio stations carried his program around the U.S., while short-wave or Armed Forces radio links sent his voice over the Atlantic. Practically any ship with the right antenna could hear Walter Winchell.

ALL THE SHIPS AT SEA

The importance of shipboard radio was first demonstrated—very dramatically—when the R.M.S. *Titanic* sank in 1912. Distress signals from the ship led to the rescue of one-third of her passengers, and news of the sinking reached the newspapers through a network of telegraphers from the rescue ships, working with land-based counterparts in Cape Race, Newfoundland, down the coasts of Canada and New England to the media mecca of New York. The Titanic disaster, which killed nearly 1500 people, occurred within earshot of the rest of the world.

Since the 1920s, no commercial vessel has gone willingly to sea without a two-way radio. In affluent countries, it is expected that fishing boats and pleasure yachts will carry a radiotelephone. Even a citizens' band radio, costing less than $100.00, has saved lives in an emergency. In 1980, a Norwegian cruise ship caught fire in the Gulf of Alaska, but because it was in continuous communication with land and with other ships, no lives were lost. The evacuation proceeded without disaster because of satellite relays.

Ship-to-shore communication is vital, but most people never see or hear it. By far the bulk of it comes from weather reports, office business between a shipping line and its fleet, changes in regulations or arrival times, and passenger services such as telegrams; most of these can be handled by a teletypewriter, which accepts telegraphic signals and prints messages on paper. Printed communications are known, collectively, as "record" traffic. There are telephone calls and other voice communication, but record carriage takes up a great deal of the time on maritime communication channels.

Unfortunately, short-wave (HF) radio such as marine radiotelephone is not always reliable, and is less and less able to handle the increasing demands: teletype, voice, facsimile, and data are clogging the airwaves among ships, ports, and offshore platforms. In most cases, only one signal can be sent at a time, and the bandwidth of the short-wave frequencies is just enough to carry one voice in one direction at a time.

International treaties limit the number of frequencies over which maritime communication can be made, and in regions where there is a lot of shipborne commerce, competition for time on those frequencies is intense. Once someone is "on," no one else can use that frequency. Short waves, in theory, carry over long distances, but natural phenomena such as disturbances in the ionosphere (which reflects HF waves) can limit shortwave propagation unexpectedly. Also, HF is fairly expensive, because it requires human involvement in setting up connections, "tweaking" the controls, etc.

In 1976, the first escape from those limitations was made with the help of a satellite network called Marisat. With one satellite over the Pacific, and another over the Atlantic (a third over the Indian Ocean was launched in 1978), Marisat carried teletype traffic to and from ships and offshore oil drilling platforms. With a footprint as big as a whole ocean, each Marisat bird was demonstrably more reliable (and could carry much more traffic) than conventional HF radio could.

Three years later, an international maritime satellite compact formed INMARSAT, which was patterned after INTELSAT, but with this difference: the Soviet Union and its allies are also participants. The INMARSAT satellites are currently used to carry maritime traffic for its signatory members.

INMARSAT birds are geostationary, and they operate on frequencies reserved for satellite maritime services by the ITU. Communications from the shores are uplinked at 6 GHz and relayed down to ships at a much lower frequency, 1.56 GHz; the ships transmit at

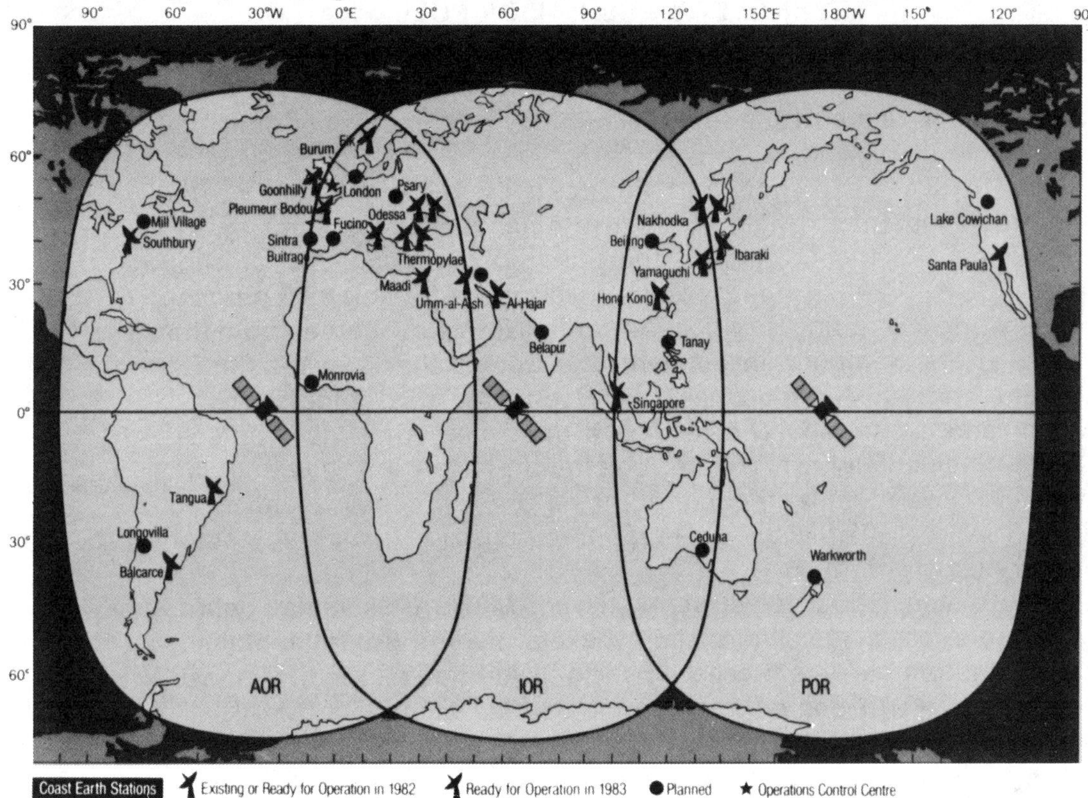

With three satellites, INMARSAT can cover nearly every possible shipping lane. (Courtesy INMARSAT)

1.6 GHz and the satellite converts that to 4 GHz for receipt on land. The difference is significant because the INTELSAT or INMARSAT earth station for 6/4 GHz traffic is the "Standard A" size—a whopping 30 meters in diameter—far too large even for an ocean liner or even an aircraft carrier to bear. Using lower frequencies between the ships and the satellites permits them to carry antennas that are only one meter wide, weighing only 25 kg. They have built-in stabilizers to counteract the ship's movement and keep them aligned with the appropriate satellite. Of course, the trade-off is that the lower frequency and smaller antenna permit the ship to use only one two-way voice channel and one two-way teletype channel at a time. Such a small system cannot possibly carry as much traffic as a larger, earthbound dish can.

TECHNOLOGY-PUSH; USER-PULL

Olof Lundberg, its director-general, says INMARSAT was created not merely because the technology was available but because users and potential users needed it. "Technology-push and user-pull," he calls it.

INMARSAT is owned and financed by telecommunications "entities" appointed by member governments. COMSAT, for example, represents the U.S.; post and telegraph companies represent many others. The 37 signatories are responsible for some 85 percent of the world's shipping. Political power in INMARSAT comes from the shares of its members' investment: the U.S. is the largest, with 23 percent; the Soviet Union has 14 percent; the U.K., nine; Norway, seven; and Japan, six. The rest have three percent or less apiece. Liberia, incidentally, which is a registration-haven for shippers who prefer to have minimal regulation, has a 0.05 percent investment in INMARSAT, yet it officially governs more ships (297) than any other country except the U.S.

More than 1500 ships currently have INMARSAT systems aboard: oil tankers, passenger liners, ice breakers, fishing boats, container, cargo, and ore carriers. Research ships in Antarctica, Mediterranean yachts, and North Sea oil-drilling rigs are also equipped. INMARSAT services include telephone, telex, data, and facsimile that are in every way comparable to the devices now penetrating offices in the major cities. "In 1982, our business increased 60 percent," says Lundberg, "and I find that both stunning and overwhelming. We expect the number and range of users to increase in the coming years."

Lundberg particularly looks forward to the day when Morse code will be abandoned. After 80 years as the backbone of emergency safety systems, hand-operated keys will be replaced by automatic machines. "The radio officer will be freed for other, more productive duties than manual watch-keeping," says Lundberg.

> In his present role, the radio officer costs the ship owner more per year than the investment in a satellite terminal. This fact alone indicates that vessels which today have to carry the mandatory Morse system might prefer INMARSAT's automatic services, with easy, immediate, global, 24-hour availability.
>
> We are looking toward development of smaller terminals, suitable for smaller boats used in the fishing industry,

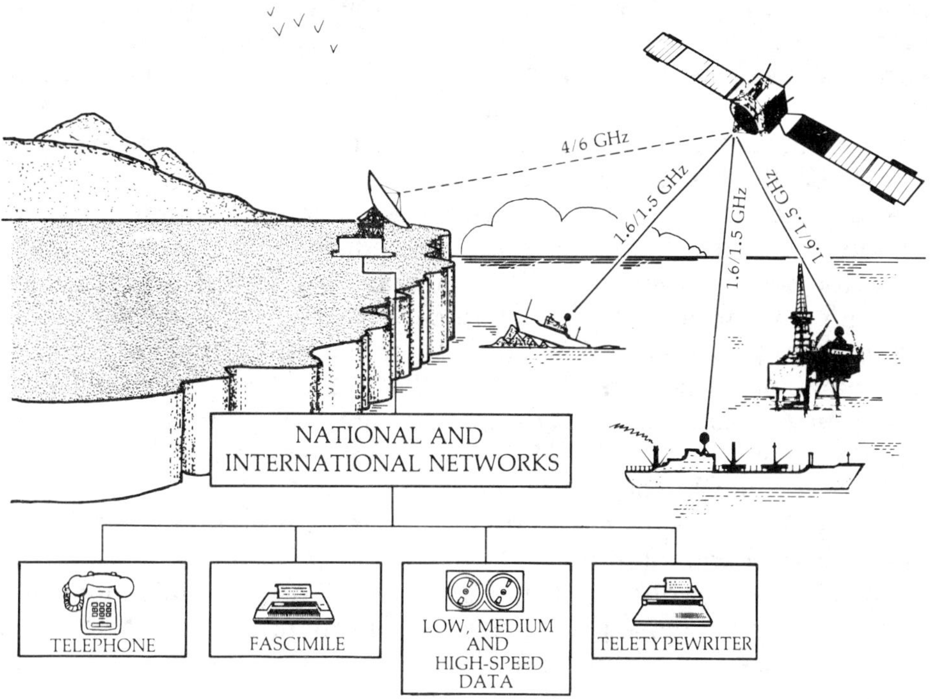

INMARSAT's system (greatly simplified) uses different frequencies for different customer's needs, but voice, data, and teletypewritten messages can be transferred among all of them.

> that would cost less than the $50,000 users currently pay for the shipboard unit. There are more than 70,000 vessels in the world above 100 gross tons, and even with our present terminals we have customers with still smaller vessels. The INMARSAT user population would be well in the range of 20,000 to 30,000 ships in the early 1990s.

INMARSAT is actively pursuing position/location technologies. Ultimately, says Lundberg, all commercial vessels will have to carry such beacons, so rescuers can home in on a ship in distress. Navigation can also be enhanced by satellites: those which loop around the earth in polar orbits (see Chapter 15). "For such future services—as they have in the past—governments will have to assume the costs," he says flatly.

Lundberg also sees INMARSAT participating in a "single, international satellite system which provides services for communications, including distress and safety, weather data relays, search and rescue, and, perhaps, even navigation. We believe we can provide services that lead to economies not only in operation but also in scale. Because the cost of satellite facilities are shared, the cost of any particular service will be lower than it might otherwise be."

When the Marisat project began, several Pacific island nations asked if they would be able to use the satellite for international telephone and record traffic, as a replacement for ATS-1, when it finally dies. Too poor to afford an INTELSAT Standard A or Standard B dish, and too small to have a population capable of generating that much traffic anyway, the island nations wanted access to some services at a reasonable price. Unfortunately, the ITU's distinctions between "land" and "maritime" mobile services are too restrictive to permit such a mixed use, and INMARSAT does not want to compete against INTELSAT. As a consultant to NASA once said, INMARSAT connection would be possible only if the Pacific islands were reclassified as "permanently anchored ships."

Still, the "maritime" bands are used for communication with offshore oil rigs. Those platforms can be as immobile as islands, so it is hard to see any technical justification for denying the island nations a share in the maritime bands. They are, after all, as utterly isolated as any ship at sea.

Chapter 9

ATS-1: PROMISES AND COMPROMISES

They are all "islands": the developing nations of Asia and Latin America, the villages of the Arctic, and the thousands of bits of land awash in the Pacific Ocean. Isolation is their common heritage.

People who need medicine, agriculture, education, and commerce struggle to bring them in over vast distances. Transportation is infrequent and sometimes dangerous; news of weather, epidemics, or opportunities cannot reach the people who need it most. Communication can help, but the appropriate technologies are not always available. And yet, for many years, the isolated people of the Pacific have had a satellite that they could practically call their own.

PEACESAT

In 1966, NASA launched its first applied-technology satellite, ATS-1, as a tool for learning communication skills and testing radio equipment. After a few years' use by the scientific community of the U.S. it was left hanging over Christmas Island, in the center of the Pacific. NASA had finished its experiments and was building more sophisticated satellites. By 1971, ATS-1 had a status similar to government surplus hardware, so NASA made it available to nonprofit groups at no cost.

ATS-1 was a repeater for radio signals: it listened to those beamed up at it, and sent them back down to receivers, amplified but unchanged. It was very high-powered, so the signals from earth could be comparatively low-powered. A village, or a school, or a hospital needed only a simple VHF radio—the kind that city police or fire departments use—and an antenna that was cheap and simple

A NASA technician examines one of the antennas aboard ATS-1 a few months before its launch. This small satellite has been in continuous operation since 1966 and may be able to run for another ten years. Its most intensive user has been the Pan-Pacific Educational and Cultural Experiment by Satellite (PEACESAT) project, which brought inexpensive two-way and conference-call services to remote Pacific islands. (Courtesy NASA)

enough for an amateur radio operator to build out of scrap parts. That's it. Emergency medical service providers in Alaska were among the first to use ATS-1, finding it more reliable than shortwave radio, particularly for reaching remote settlements.

In 1971, a group of users set up an international consortium for Pan-Pacific Educational and Cultural Experiments by Satellite. As its acronym suggests, PEACESAT offered equality of opportunity—and price—to every participating country, regardless of size.

PEACESAT's principal American organizer was University of

Hawaii professor John Bystrom. From the first, his satellite projects were designed to help people, and not simply to prove that a thing could be done. He had noticed that the University's library used inter-island mail for book orders from the Hilo campus, on the island of Hawaii. Using PEACESAT, he linked Hilo with the Honolulu campus, on Oahu, and when the library transmitted book orders by facsimile, it cut delivery time by days. Not long after, he brought about the first satellite-borne course for college credit: students on three islands "met" in an electronic classroom.

Bystrom claimed in 1975 that if PEACESAT were adequately funded, it would, in his words, "provide communication resources to remote locations [that] in quality, flexibility and range will be *superior* to those enjoyed today by the top executives of the most advanced countries" [emphasis his].

In the 1970s, Bystrom's enthusiasm and optimism were boundless.

> The needs of the world's people for education, health and community services justify the development of a new kind of telecommunication system [which] will not fit precisely into existing classifications . . . but will have qualities of all of them. The PEACESAT project tests a work-oriented system. It is not to be confused with the mass media and with entertainment-oriented communication, although it may assist the mass media in such tasks as the distribution of news. It is not designed to accelerate the rise of expectations among the world's people but rather to assist those who carry the responsibilities for meeting those rising demands.

Bystrom's closest colleagues were teachers at Wellington Polytechnic, in New Zealand, and at the University of the South Pacific, in Fiji. They experimented with nearly every kind of low-budget, narrow-band communication technology, including voice, teletype, and facsimile; they linked the satellite to local telephone networks, so users did not have to gather, physically, at the PEACESAT terminal sites. By the mid-1970s, there were terminals in Honolulu, Kahului, and Hilo (Hawaii), Wellington (N.Z.), Suva (Fiji), Lae and Port Moresby (Papua New Guinea), Saipan (Trust Territory), Noumea (New Caledonia), Niue, American Samoa, Anchorage (Alaska), the University of California at Santa Cruz, Vila (New Hebrides), Rarotonga (Cook Islands), Honiara (British Solomon Islands), Tarawa (Kiribati), and the Kingdom of Tonga.

the air at a certain time for a conference call, with one of them serving as the coordinator of that session. The satellite would accept one uplink signal at a time, and broadcast it down over the entire footprint from Washington D.C. to Australia. Before anyone else could respond, the speaker had to say "Over" or "Back to you, Saipan" and switch off his or her transmitter, but that was not seen as a limitation. Many small countries have more shortwave radios than they have telephones anyway, and the protocol was already familiar; besides, any inconvenience was vastly offset by the enormous footprint and the low cost of entry into the system.

In many of the island nations, the inexpensive telephone-like communication appealed to radio amateurs, academic professionals, and government officers. Few of them had direct access to submarine cables for voice traffic, and those that did paid what seemed to be high rates for short phone calls. Bystrom's hope for a low-cost network equal to or better than those available to heads of state seemed close indeed.

With its conference-call format, PEACESAT helped keep doctors and nurses up to date; it brought agricultural extension agents the latest techniques, and it overcame the prohibitively vast distances that had kept the islanders from discussing, face-to-face, issues of mutual concern, such as tourism, fishing rights, and regional economic planning. The University of the South Pacific used ATS-1 to teach extension courses far beyond its Fiji campus, and the World Health Organization brought post-graduate education to medical professionals.

An outbreak of dengue fever was checked because medical know-how got from Honolulu to Suva in time. The information exchanged *could* have been routed through trans-Pacific telephone cables, but the costs were considered prohibitive at the time. Less-critical material probably would never have been carried by telephone at all, including conferences on "Problems of Urban Renewal," "Brownie [Scout] and Girl Guide Leader Training," "Organic Waste Digesters," and "Destruction of Coral Reefs by the Crown-of-Thorns Starfish."

What kept the terminal cost low—under a thousand dollars, in some cases—was NASA's decision to put the power in the sky instead of on the ground. The trade-off comes not only in design technology but in *concept.* On one hand is a low-powered "mirror" serving large, high-powered earth stations which can gather up and concentrate many signals at once, such as those generated by a major city. On the other hand is a stronger satellite to amplify weak signals from small terminals, when the locations of those signals are widely scattered, or their appearances on the air unpredictable. ATS-1 took the second approach and built the power into the bird so that the earth stations could be small, cheap, and even portable.

Papua New Guinea—to cite just one example—was a beneficiary of PEACESAT's "appropriate technology." The PNG University of Technology bought and assembled a 100-watt transmitter and an antenna for just US$1100. "A 10-watt ground station would be ideal for village-to-village communication," said one instructor, "because of its low drain on available energy sources and its portability. The cost of a 10-watt prototype is about $500." Another $500, he said, could provide an air-conditioned studio. There is a photograph in PEACESAT's scrapbook of an entire earth station—generator and all—loaded into an islander's canoe; another shows one hauled by jeep into the interior of Papua New Guinea.

HUMAN FACTORS IN SATELLITE COMMUNICATION

Despite its technical success, PEACESAT was slow to catch on with people. Kit Porter, an educational administrator in the Northern Marianas Islands, had used the PEACESAT network to discuss education issues with her colleagues abroad but, she says, "what surprised me was people's hesitancy to use the equipment and talk. In a place where phone connections are undependable and not existent between most Micronesian islands, PEACESAT seemed to be an underused educational resource." It is also possible that, because telephones are rare, the Micronesians did not have enough experience using them to feel comfortable talking over great distances.

Jim Dator, a University of Hawaii professor, helped Porter create a graduate-level political science course ("The Politics of the Future"). With Dator visiting on Ponape, the students on Saipan, and Porter temporarily in Portland, OR (connected by telephone with a PEACESAT station at Santa Cruz, CA), the class "met" for the first time. "There were negative elements to satellite use," Porter admits. "The transmission was not always clear, and sometimes sections had to be repeated because of the deteriorating orbit pattern of ATS-1. More coordination than had been expected was needed on the Saipan end to assist students, most of whom were working in a second language. These were minor, however, compared to the benefit of the learning experience."

For the fifth session Dator was in Alaska, and the following week he was in Florida (patched by phone through Denver, CO), but by the end of the course the difference between North American communication networks and the islands' makeshift transceivers grew wider; on Palau the transmitter failed and they could only receive signals. Porter reached these conclusions: "The frustrating experience [on Palau] leads people to give up on the system; there must be a course coordinator on site to follow up, and to encourage students; and the course must fill a need felt by all the participants." She wants to try again.

The University of the South Pacific was one of the first PEACESAT users, and one of the first to break away from that group, citing increased internal communication needs. With nine ATS-1 hours a week of its own, USP activated terminals in Niue, the Cook and Solomon Islands, Kiribati, and Western Samoa. The World Health Organization, International Labor Organization, and several agricultural groups have since used USPNet. Jay Miller, of USP's Extension Services, says that people in the region are accepting—and using—this new mode of communications. "Despite the criticism that a satellite network primarily reaches the elite, there is little doubt, in the opinion of the users, that it has contributed to getting new ideas flowing around the Pacific," he said.

USP has expressed interest in using ATS-1's voice-grade circuit for data; John Southworth has already demonstrated that capability. A University of Hawaii instructor, and a pioneer in computer communication, Southworth set up a PEACESAT session between Honolulu, California and Western Samoa, with a "host" computer in New Jersey's Electronic Information Exchange System (EIES) and off-the-shelf microcomputer equipment at the Pacific sites. EIES can link dissimilar computers together, and can cope with the quarter-second delay introduced by the satellite link.

The data-teleconference broke no new ground but it reinforced Southworth's experiences over the previous two years. In 1979, he experimented with slow-scan television and facsimile, and in mid-1980, he connected PEACESAT terminals to Control Data Corporation's huge PLATO educational computer system. He called it a "multi-mode node," and proved that electronically incompatible signals, such as analog voice and digital computer data, *can* be mixed into one conference over ATS-1. At an international symposium, Southworth presented a paper on the subject which he had written in collaboration with other researchers: it was composed on

PLATO terminals (over the Control Data network) and its text had been transferred over ATS-1 from PLATO to a Xerox® word processor through a Texas Instruments portable computer terminal!

Southworth's most ambitious computer conferencing project used ATS-1 to link microcomputers in various locations. In 1981, a software program and some data for it were transmitted by John Lovell at USP, using a modified Apple® microcomputer, to the PEACESAT terminal in Honolulu where it was captured by the TI portable terminal. Intermittent bursts of static caused several parts of the transmission to have a high error content, but the overall error rate was less than 0.4 percent, Southworth reported. The 17 transmission errors in a 105-line program sent from Fiji to Hawaii (in about 3 minutes) were quickly corrected by comparing the texts from just two program transmissions. The importance of this kind of "gee-whiz" experimentation was that it proved ATS-1 could make the leap from the world of voice communications to the realm of microcomputers. It had, after all, been built for power and simplicity in the first place. Technologically and economically it was a success.

PEACESAT'S CRITICS SPEAK

Socially and politically, however, the promise of satellite communications has not fared so well. Christopher Plant, of Simon Fraser University, served as a PEACESAT terminal manager in Vancouver, B.C., for several months, and delivered a surprisingly critical report to his colleagues.

> It seems that interested parties are asked to infer from the existence of a technically operating system in the Pacific Islands that it is automatically of benefit to the development of those islands. . . . Regarding PEACESAT's asserted social value, there exists only anecdotal evidence concerning users and uses of the system [including] use by medical authorities trying to control epidemics. . . . Accounts of such isolated experiments are impressive; at the same time they are not representative.

Plant undertook quantitative analysis of PEACESAT station logs and came to some unhappy conclusions. "Metropolitan locations" (i.e., developed countries or urban areas) used PEACESAT nearly twice as often as "Pacific Island" (i.e., rural or remote) sites did. The number of Caucasian participants exceeded the number of non-

Caucasian participants by a ratio of almost two to one; they used nearly four times as much air time, and accounted for more than 90 percent of all session chairmen. "Only 1.8 percent of all exchanges were initiated by Pacific Island people," Plant said, "despite their terminals outnumbering Metropolitan terminals two to one." Plant saw PEACESAT as "dominated" by Honolulu and Wellington, and attributed its lack of penetration into island life to its having been, as he said,

> mistakenly regarded by its organizers as a technology, in the narrow, traditional sense of the term. . . . [They saw it] as meaning primarily the hardware, rather than as a composite of the institutional, administrative and inter-personal relations that necessarily accompany the hardware. . . . PEACESAT per se cannot be considered an appropriate technology for the Pacific Islands, and the PEACESAT project has failed to provide the research necessary to show that a similar system can be appropriate in the future.

John Bystrom admitted, at a 1981 PEACESAT users' meeting in Wellington, that "there is a problem of people who are overdominating in discussion." But he countered that with observations of his own.

> The system is designed to leave communication to the individuals involved in the matter under discussion, not to delegate it to media professionals. Terminals owned and operated by indigenous people are needed. . . . It takes time for most people to break down inhibitions to "going on the air." Often, when an individual hears others speaking about experiences that are common to both, he is tempted to speak. Probably every participant should go through preparatory exposure and training for best results but resources have not permitted enough work in this area even after ten years.

Historical precedents for the use of high technology in developing countries "are not particularly encouraging," says Craig Ritter, director of a two-way television program in the Irvine, CA, Unified School District. "This is due in large measure to the fact that most aid programs have addressed themselves to the needs of a small, powerful elite, those who may have been educated abroad and who already possess the value systems and motivations of developed societies."

Ritter studied various telecommunication experiments in developing countries and disputed their conclusions that two-way audio is an "adequate" vehicle for instruction and that one-way video with audio return is "sufficient." Lack of prior experience with interactive systems other than telephone," Ritter says, prevented them from "reaching a point where they considered the technology to be part of the means of communication rather than artifice."

As most countries have some experience with broadcast television, Ritter advocates a plunge directly into television, on grounds that one-way systems "do not allow the development of interpersonal relationships among teachers and students." But television requires so much bandwidth that the cost of facilities, equipment, and microwave or satellite transmission is enormous. (ATS-1 cannot carry tv.) Institutional barriers exist in many nations forbidding reception of neighboring countries' television programs, and severely restricting the flow of computer data.

AN ARCTIC SUCCESS STORY

PEACESAT-style audio conferencing, with some inexpensive enhancements such as facsimile, are realistic alternatives to the silence of distance. Alaskans live so far from one another that only telecommunication can link them, especially during the winter. Farsighted legislators established an audio teleconferencing network there in 1978, a dedicated four-wire telephone circuit for citizens to participate in decision-making. Sioux Plummer, who coordinates the network for the state government, admits "it exists because there is a well-defined need to provide people with this type of communication, and because those same people like using it. The majority are repeat users, including persons who testify in public hearings, and those who attend workshops or training sessions. Established protocols and policies are now good ones, leading to a successful format."

The system works, she says, "because it is constantly used. Daily, in between scheduled teleconferences, there is administrative chatter and information exchange among the 18 teleconference offices around the state. It is therefore constantly monitored for problems such as adverse line noise or interference which are reported to Alascom, the system operator, immediately." During the legislative session there have been as many as five different teleconferences in a single day.

It costs the State of Alaska $16,000 a month to reach 75 percent

Inside PEACESAT: *One reason why ATS-1 has operated for so many years is the careful attention which it received in the laboratory prior to launch. Most of its components have backups to take over if the primary ones fail. (Courtesy NASA)*

of the population. Compared with air travel over the arctic wilderness, and the time that it consumes, Plummer says, "citizens often report that they would not have taken the time away from their job or family to testify in a public hearing had the network not been so convenient and efficient." It will not remain what she calls a "whoop'n'holler" network for long. Sophisticated selective signaling will notify terminals when a conference is in session—particularly useful for the most distant users: Alaska's congressional delegation in Washington, D.C. Robert Walp was the state's telecommunications director when the system was installed. In his view, years later, "it works because it has to."

PEACESAT was, like the Peace Corps, an idealistic endeavor. People gave their time and talents unselfishly to help other people. That ATS-1 is still available for low-cost teleconferencing is testimony to the goodness inherent in the idea and in the people who made it happen. (It also speaks well of the NASA engineers who built ATS-1 so well that it has endured far longer than any of them ever imagined.) In the future, however, the Pacific countries will not be able to rely on that kind of American largesse. The next chapter explores some of the hard choices they will have to make.

Chapter 10

SATELLITES FOR THE PACIFIC

Satellites would appear to be the best way for Pacific region countries to communicate. Cables can connect only the largest, most populous islands, and the nations around the rim—with few exceptions—are already industrialized.

THE OUTLOOK FOR THE ISLANDS

The most convenient alternative to satellite communication is high-frequency (HF) radio—popularly known as "shortwave" radio. It came into its own during World War II and has continued as the most nearly ubiquitous communication tool in the Pacific. Practically every island in every group has a two-way radio and someone who at least knows how to use it, if not fix it. HF is very well suited for voice communication because, even though the signals can fade or be distorted, human ears are very "forgiving" and can interpolate meaning from garbled transmissions. HF is not suitable, however, for data, facsimile, video, or other more demanding services.

New technological improvements have kept HF cost-effective. Some automatically find a transmission frequency that is free from interference, and others "scan" all received signals to isolate those which are desired (as city police radio "scanners" do, locking onto emergency calls). But HF is expensive, in absolute terms, and not much cheaper than some satellite-borne alternatives.

It would be useful for social and economic development, and to further the Pacific countries' goals of becoming full participants in world affairs, if satellite communication were both available and affordable. Unfortunately, that may not happen easily. All the alternatives are difficult.

PEACESAT IS CHEAP

ATS-1 has a global (or at least transoceanic) footprint. A signal can be sent to or received from ATS-1 virtually anywhere over one-third of the world. By contrast, the INTELSAT network uses large dish-shaped antennas on the ground to concentrate the transmissions, and to collect the faint signals from the satellite. Though its satellites are larger than ATS-1, their on-board power is divided into many more separate but weaker transponders, to raise the total number of two-way circuits. Each transponder's footprint is kept small to concentrate its high frequency signals. Each earth station costs at least $100,000 and requires a support team of trained electronics technicians.

INTELSAT carries thousands of simultaneous telephone conversations and dozens of television broadcasts worldwide; ATS-1 carries only a few phone calls. That is typical of the progress in engineering since 1966, but it has placed the island nations at an extreme disadvantage. Without large industries, or outposts of transnational corporations, they do not have nearly enough of the domestic or international telephone traffic needed to support the cost of installing an INTELSAT earth station. A few have found financing through the World Bank and other lending institutions, but the tiny places—the ones that can afford only the low-cost PEACESAT terminals—may not be able to join INTELSAT in this century, if ever.

The ATS-1 concept, with low-powered earth stations using a high-powered satellite transponder, is one that many developing nations would like to apply to new satellite systems. But the industrial countries which launch and pay for them can save weight and cost by leaving the high-energy antennas on earth and multiplying the number of weak antennas aloft.

BUT PEACESAT NEEDS CONSTANT ATTENTION

ATS-1 has been working for over 17 years, *nearly three times its original life-expectancy!* Still, it requires constant telemetry, and every slight adjustment to its orbit position, its rate of spin, or its orientation vis-a-vis the earth, uses up more and more of its station-keeping fuel (hydrazine, a catalytic propellant). By 1983, ATS-1 had been moved to a more "stable" location at 195° W, but it cannot live forever, and there is no comparable system to take its place.

NASA scientist Louis Ippolito was, frankly, surprised at its tenac-

Installing a COMSAT Earth Station—*These photos show the installation of a COMSAT earth station in West Virginia–one of three located there. They are used to communicate with INTELSAT satellites. The size of the operation, with its demand for heavy equipment, is an indication of the reason why third-world countries may be unable to install very many comparable dishes. (Courtesy COMSAT)*

ity. "When we moved ATS-1, the amount of hydrazine used was less than we expected, so we are now projecting an 'indefinite' life for it—certainly ten more years." Plans call for NASA to cede responsibility for the telemetry of ATS-1 to the University of Hawaii in 1984. A similar arrangement was made with the University of Miami (Florida) to monitor ATS-3, which—like ATS-1—is used by nonprofit groups and state governments for emergency medical services, search-and-rescue, and other humanitarian purposes.

The university would pay two people a year ("one and a half man/years," in government parlance) to operate the telemetry equipment. Says Ippolito, "It requires the computing power of a DEC PDP-11 [an industry-standard minicomputer]—which we will supply—to monitor the satellite. Its on-board clock has to be reset once a day, and there are other commands which have to be radioed up to make routine changes. About once every six months, the operators will have to transmit telemetry sequences for station-keeping."

THE PERSISTENCE OF THE "ATS-1 MENTALITY"

Sir Thomas Davis, prime minister of the Cook Islands, regrets the loss of cheap circuits. Ten percent of his people have used PEACESAT, and through it the Cook Islands have organized and jointly conducted with overseas scientists much more scientific and medical research than would otherwise have been possible. With solar (photovoltaic) batteries, even the farthest islands in that group can receive news broadcasts, and participate in political decision-making. Davis is a ham operator with a shortwave *and* a PEACESAT radio on his desk. The eventual loss of ATS-1 is, for him, a personal loss.

But Marcel Perras, director of business planning for INTELSAT, is critical of what he calls the "ATS-1 mentality" that is "unconstrained" by the economics of modern satellite development. "They [ATS-1 users] are looking for a dependable system, with continuity, to expand their community of interests over the long term. But that will not happen," he insists, "unless communications development is put on a commercial basis."

According to Robert Lovell, director of NASA's communications division, NASA's budget does not include money for research and development of low-cost earth stations, except those used with direct-broadcast television experiments. Those antennas will allow people only to receive signals, and not to transmit or share anything of their own, as the ATS-1 system does.

"Several major problems remain to be solved," says Heather Hudson. She was a planner at the Academy for Educational Development, in Washington D.C., before joining the faculty of the University of Texas. "First, the smallest political entities must also have reliable communication— *sine qua non* for participation in international affairs and running an independent government. But even with a satellite station in each country or semi-autonomous region, the problem is far from solved. The Pacific contains thousands of scattered islands; most nations are composed of several inhabited islands. Many of these are small, and too distant to be linked reliably by terrestrial radio repeaters." The cost of undersea cable would be prohibitive in almost all cases.

"Another issue has to do with the type of service ATS-1 has provided," Hudson says. "It is a shared conference facility available at negligible cost. Conferencing has proven to be a useful tool in the Pacific. The present users could not afford to use a conferencing service if current international rates for point-to-point communication were applied."

PACIFIC SATELLITES OF THE FUTURE

Since ATS-1 could, theoretically, fail at any time, its users hope that some other satellite will be available to take its place. However, a "surplus" satellite that could give the poorest nations cheap and reliable voice communication, as ATS-1 can, does not now exist. The people of the Pacific will either have to buy a new one, attach themselves to some existing service, or persuade a carrier to make transponder time available at a lower price than ever before.

In a report prepared for NASA, the Public Service Satellite Consortium (PSSC) evaluated the options, and concluded that neither the construction of a dedicated satellite, nor the relocation and/or recycling of an existing satellite was suitable or practical.

Build New Satellites?

One satellite that can be used by people with low-cost terminals is the ARABSAT design, built for a Middle East consortium by Ford Aerospace. It uses frequencies around 2.5 GHz (the S-band), so it needs only modest power; it has a broad footprint, and its simple earth stations—though fairly large—can be built by relatively unskilled workers. The S-band is already allocated to both "fixed" and "mobile" communication services (i.e., for both antennas which are permanent and those which can be moved), which is vital to villages where there is no place for a fixed antenna.

Unfortunately, the Pacific countries probably can't afford to buy a new satellite, as the oil-rich Arab states can. "The cost and schedule for the design of a special satellite of this class," said the report, "would be prohibitive in a region where resources are so few and economies so fragile."

PSSC reported it heard reported "rumors" that Hughes Aircraft Co. had enough components on hand to build a new satellite based on the design of the early WESTAR, ANIK, and PALAPA—a 12-transponder bird. Because it was an unbuilt spare for Western Union, WU still owns it and may not wish to sell it. If it were for sale, it would probably cost at least as much as the price of building a brand new, 24-transponder satellite, and in any case it would have to be built without the assurance that a spare was available. That's considered unwise because launches sometimes fail.

A Japanese research team proposed, in 1983, that a special satellite could be built to serve the Pacific islands. It would offer a large footprint in C-band for the countries with only a little traffic, and a few narrow spot beams in K-band to give the urban centers digital services. While it was technically appropriate to the region, and its customer potential was well-researched, it lacked clear governmental or financial commitment. Japan would like to be the technology leader in the Pacific, but it is unlikely to create a satellite system of oceanic magnitude out of altruism alone.

Recycle Old Satellites?

PALAPA A1 and A2 are Indonesia's first domestic satellites. They are spin-stabilized Hughes birds, carrying 12 transponders each, and are due for replacement by 24-transponder satellites in the mid-1980s. In theory, they could be moved somewhat to the east to cover the Pacific basin.

But—being used—they may not be entirely reliable, and if they run out of stationkeeping fuel they will drift toward one of the stable points in the geostationary orbit—a kind of "graveyard" for dead satellites near 106° W. While the PALAPA birds are more or less "available" and are known to work, experience with their sisters—

ANIK and WESTAR—shows that several transponders have already failed or can be expected to fail by the end of PALAPA's normal design-life.

WESTARs I, II, and III are also available, but they, too, are old and tired. If they were moved to the mid-Pacific they could be inverted so their antennas pointed to the southern hemisphere, as PALAPA's do. However, since WESTAR was designed to reach the higher northern latitudes, where the U.S. is, it would cover only the higher southern latitudes well, and probably miss the equatorial and tropical areas where most of the Pacific islands are. This would be a problem for reconfiguring any North American satellite.

It is possible, however that the U.S. could recycle another experimental satellite as it did with ATS-1. That would be one of the Tracking and Data Relay Satellites (TDRS) that monitors low-earth orbit birds from its geostationary position. It has a stronger signal than ATS-1, but a narrower footprint—not Pacific-wide. But TDRS has not yet outlived its usefulness; indeed, the first of three was launched in 1983 and failed to reach the geostationary arc. Even when it is properly aligned and begins full operation, it will merely be taking over from ground tracking stations for the next few years. TDRS is by no means "surplus," as ATS-1 was when it was given over to noncommercial users. And the TDRS design is very complex. It's also known as "Advanced WESTAR" because one of its sisters is due to join the commercial service, and it features movable antennas and transponders that can serve multiple frequencies. NASA scientist Louis Ippolito says conversion of TDRS "does not appear viable for early use, certainly not before 1984 or 1985," but he does not rule it out.

Ippolito sees greater promise in lowering the cost of earth stations for existing satellites. NASA supports research and development of portable, battery-powered terminals, such as one made by Westinghouse for accessing ATS-3, which costs about $15,000. But light weight has penalties: Ippolito calls one contraption he has seen, "a little rickety."

The Intelsat Alternative

The PSSC study concluded that some use of INTELSAT would be "the most available, workable option for the immediate future." Heather Hudson is convinced that

> the satellite picture in the Pacific is rosier than planners anticipated five years ago. Within three years it's likely that only

> the smallest entities, such as Niue, Tokelau and Tuvalu, will not likely have INTELSAT stations. In the Western Pacific [i.e., in the Trust Territory], the U.S. plans to install earth stations in each regional administrative center. The station in Saipan is already operational.

GTE has long recommended that small countries pool their resources and share the cost of a transponder; they would generate sufficient traffic by linking their regional telephone systems over cables and HF radio to GTE's COMSAT earth station in Hawaii.

In 1981, INTELSAT proposed that small countries lease one-quarter of a transponder, for US$200,000 a year, and use solar-battery powered, 5-meter earth stations costing approximately $150,000 apiece. In 1983, INTELSAT began approving $40,000 dishes, and expressed a willingness to help small countries buy or build them. Joseph Pelton, one of INTELSAT's leading spokesmen, admits that not every village in a developing country could have a dish; rather, outlying villages would feed telephone signals to the "node" village with the dish, through HF or VHF radio units (also solar-battery powered) that would cost about $2500 each. Pelton said that the $200,000 annual lease for the space segment could be spread over 50 voice channels, and cost only about $4000 per circuit. If each circuit's traffic averaged 300 minutes a day, for 300 days a year, he said, the users' cost (or the government's) would be a little over four cents per minute. But that assumes that enough earth stations or radio feeder equipment could be amortized as well. And even the new, smaller dishes, called Standard E, for thin-route service, require professional maintenance.

Despite any reduction in the cost of earth stations, Mr. Perras says that transponder time will still be expensive. "INTELSAT has to provide the same service for the same price worldwide, and any cross-subsidization is already embedded in the structure. It is the high density of transatlantic calls that keeps rates low in the Pacific."

Sir Thomas Davis still hopes that the U.S. or some other industrial country will somehow build high-powered satellites to serve small, inexpensive terminals. "For Pacific-wide conferencing, with low traffic density, where the communications are to aid development," he hopes for "low cost terminals that can stand up for years on an isolated island without air conditioning . . . for less than $1000."

It may, officially, have "ten years" to go, but powerful little ATS-1 *is* dying, and Pacific island nations are searching the skies, waiting for a new one that will probably never come. The era of American largesse is ending, and our cast-off technology may never again be as well suited to the needs of developing countries as it was a decade ago. As ATS-1 wobbles toward its silent fate, and the powerful nations turn away from an era of unselfish giving, the prime minister's plea will not be answered. It is probably a cold comfort that the Cook Islands, Samoa, Tonga, Fiji, Kiribati, Niue, the Solomon Islands, and many other tiny bits of land in the Pacific stood—for a decade—as equal partners with the U.S., as pioneers in space.

Section 4

HOW TO MAKE MONEY IN SPACE

"The important question is not: 'What will it cost?' It always costs too much. The correct question is: 'Will it make money?' "

—G. Harry Stine, communications consultant

Chapter 11

VERY HIGH FINANCE

OPPORTUNITY AND OPPORTUNITY COSTS

Outer space, in the 1980s and '90s, will be the site of intense competition for technology and for capital to exploit those technologies. No one can now predict the size of the market for—among other things—zero-gravity, high-vacuum manufacturing. Pharmaceutical, geochemical, and biochemical engineers are launching experiments by Shuttle now, in hopes of finding a product or a process that will be worth more when coming down from orbit than its raw materials were going up.

But using space for communication is already paying off, despite the very high "opportunity cost" of getting in on the business early. As G. Harry Stine, a widely published consultant, puts it, "Frequent and easy access to earth orbital space, coupled with projected decreasing transportation costs, promises to open up a multimillion dollar series of markets in the communication/information area. . . . It is a capital-intensive program, and this bothers some people who seem to forget the incredible capital investment in existing terrestrial systems that we already take for granted."

As a business, space communication has so far been operated by the very largest national and transnational entities, but that will no longer be so. The entry fees are still relatively high, but they are lower, proportionally, than they have ever been. The price of computing power drops by ten percent or so each year; the price of long-distance communication drops at roughly the same rate, for the same reason: technological improvements. Small computer companies now find it easier than ever before to attract venture capital. The successful ones are those with innovative products that are both reliable and popular, and which fill previously unfilled niches in the marketplace. Now, small communication companies are learning from the computer sector: they are identifying vacant places in the market, preparing sophisticated business plans, and hiring professional managers. The venture capital firms are taking notice.

To understand the future of space businesses, it helps to look at the background against which they are maturing. The most successful entrepreneurs in space will be those with the broadest vision: those able to bring specialists together to create—in synergy—a whole greater than the sum of its parts.

FOR THE INFORMATION-RICH

The role of economics in telecommunications is not always apparent. A widespread impression is that "someone" sets the rates while "someone else" pays them. In developed countries such as the U.S., users tend to set their own rates. They act through representatives who are either elected directly or are appointed by elected officials. The "public utility commissioners," or whatever they are called, are almost always *local* people who will have to live with the rates they set.

The supply of services and the demands for them are compared by these representatives. Where a "natural" monopoly exists (a single telephone company, for example) it is customary to build a guaranteed rate of return on investment into the rate structure. If it is ten percent, then the rates users pay will be just sufficient to give the utility a ten percent profit, assuming projections of demand are accurate. Shortfalls are usually made up with rate hikes; excess profits must be reinvested or returned to ratepayers in the form of new services at no additional cost, or by delaying other rate hikes.

That procedure works because the end-users of a telecommunication system have been given control over the system by a federal government committed to promoting local control. Decisions reached through a representative democratic process, in general, convey a sense of "fairness" to all parties, even to those who are overruled or outvoted, and that has been the thrust of American public policy for 200 years.

American common carriers are not permitted to own or control the traffic on their systems, whether those systems are canals, railroads, or communication networks. AT&T, for example, cannot have a financial interest in the network television programs which it carries over its long-lines or satellites. That way, a common carrier will not tend to favor any customer over any other.

Naturally, there are volume discounts for the largest customers, and other incentives for attracting business, but the rules of interstate commerce require a common carrier to establish uniform rates for every kind of service it offers, and it must not deviate from those prices without a nod from the appropriate regulatory authority.

Prices for satellite services are not graven in stone. Competitive pressures force most carriers to change them at least once a year. New technologies, as they come onto the market, also affect the carriers' cost and the customers' prices. This will continue.

Nonetheless, it is useful to look at a few tariffs, published in 1982, as examples of how much a user actually pays to have access to satellite communication services in the U.S. The rates listed here are the published (and FCC-approved) prices for doing certain specific things over a satellite transponder.

American Satellite Co., of Rockville, MD, offers private voice and data circuits between or among major U.S. metropolitan areas. ASC encourages short-term experimentation by offering to cancel service after only 30 days' notice, but it also offers a discount if a company signs on for a minimum of one year. The all-digital network is relayed over transponders leased and operated by ASC but physically located on Western Union's Westar satellites.

Between Washington D.C. and San Francisco, for example, ASC's customers pay $1095 per circuit per month with the right to cancel in 30 days. The full-year option for the same circuit costs $875 a month. Identical rates apply between New York and Los Angeles, but the rates are lower over shorter distances. Between Atlanta and Dallas/Ft. Worth, the 30-day option rate is $820, the same as that for Chicago-Houston or Chicago-Atlanta.

STARNET Corporation, of San Diego, CA, offers private-line telephone service among the cities of New York, Los Angeles, Atlanta, Chicago, Houston, San Francisco, and Denver. STARNET imposes a one-time installation charge of $350 per circuit, payable in advance, and requires a customer to lease at least two circuits per city served, at a minimum of $600 apiece per month for "short period" service (8 a.m.–5 p.m., Mon.–Fri.).

After that, customers pay by the minute for intercity connections: among the principal cities (listed previously), per-minute phone rates are $0.18, $0.15, and $0.14 for day, evening, and night/weekend times, respectively. If additional cities have to be served, STARNET adds in the cost of AT&T Long Lines tariffs and the combined rates go up to $0.3595, $0.2778, and $0.1899, with volume discounts after 40 hours and 80 hours per month.

Bonneville Satellite Corporation, of Salt Lake City, offers video circuits for teleconferencing and program distribution through its full-time lease on a Westar transponder. A typical user might be a small-town tv station that wants to carry the local basketball team's game in another city. The program will go by terrestrial microwave to the nearest BSC uplink facility, then it will bounce off Westar to the downlink dish closest to the customer and be microwaved to the tv station for broadcast. Other tv stations might use BSC to relay their Washington, D.C. bureau's reports for the nightly news.

These *ad hoc* networks can be temporary (the basketball game) or permanent (the news feed). If the customer simply wants to send a prerecorded or live program to another city during prime time (4 p.m.–2 a.m. EST Mon.–Fri.), he or she pays $450 an hour. During "Early Bird" hours, the price is $300 an hour. Shorter programs cost proportionally less; the tariff is built so that charges are made for the first half-hour and for increments of a quarter-hour after that.

If the customer cannot feed the signal directly to BSC's facilities, BSC adds a per-channel surcharge of $125 an hour for the uplink and $75 an hour for the downlink connections. All the rates are lower for heavy volume users.

BSC also offers a facility in Washington, D.C. for transmitting news feeds and originating teleconferences; the rate includes the Westar space segment, local (D.C.) microwave and uplink service, and the live transmission or tape playback equipment. A one-way feed, for example, costs $375 for a 30-minute live show with taped inserts. Note that that is for the feed; the customer, of course, pays for producing the program and for receiving it at its destination.

These tariffs are not a complete list of possible services, but they show two things: the *way* that rates are computed varies widely, and the *price* that is paid is set according to what the user needs. A savvy manager will shop around for satellite services with the same care that he or she exercises in shopping for a telephone switchboard, a computer, or office space.

FOR THE INFORMATION-HUNGRY

But in many countries, particularly those which do not have participatory or representative democratic institutions, the procedures are vastly different. Moreover, in developing countries, rates can almost never be set high enough to pay for the communication systems in use; supply-and-demand logic fails. These are often countries whose economies are fragile, whose people may be living in or near poverty, and whose wealthy citizens are not inclined to reinvest their capital in their own nation.

"Investments in telephone communications by developing countries are usually made on the basis of a faith that the benefits achieved by the developed nations will, somehow, also be available to the developing nations." This, says Herbert Dordick of UCLA's Annenberg School of Communications, leads to a dilemma: "The question of benefits is posed in terms of communica*tions* rather than in terms of communica*ting*." Decisions are reached, he says, without considering "what went on before the call was made and what goes on after the call is completed."

The need to communicate is universal. The oral traditions of most primitive tribes were "rich in information," according to William Melody, of Simon Fraser University, Canada. "I suspect that the most significant change between advanced society today and the oral tradition of the Greek city-state, still practiced by some native cultures today, is not in the role of information in society but in the way that information processes have been institutionalized: the commoditization of information, and its sale through markets."

The high cost of making "new" information and the relatively low cost of replicating it creates special problems, Melody says, as industrial nations sell their hardware and software abroad more cheaply than the purchasing country can produce them indigenously. Even in developed countries, the imbalance is striking. "Canada's mass media is inundated with American information," he says, which puts Canada at a competitive disadvantage. "The situation reached its most ironic extreme recently when, while Canadian industry and government representatives were touring the world trying to sell satellite and related equipment and expertise, hundreds of Canadian rural communities—and some not so rural—were purchasing satellite receiving stations from U.S. firms to access U.S. satellites in order to receive U.S. television programs."

Asian professionals and government representatives were surveyed by the East-West Resource Systems Institute, of Honolulu. They spoke of obstacles to supply, and of limitations on demand, that face developing countries. At issue was data communication, specifically the trans-border flow of computer information that national or regional planners need to make their decisions. According to the findings of the report, data communication cannot be easily accomplished where there is both a lack of terminal equipment and a corresponding absence of digital networks. That was obvious—even local telephone service in most developing countries is inadequate.

But "most serious of all," the report said, "there are bureaucratic institutional constraints [on new services] often having to do with government permission," and an "even longer list [of constraints]" which limit local demand. Some needs can be met locally, as computer power decentralizes and miniaturizes. System promoters commonly "oversell" network capacity, further complicating the system from the user's perspective and introducing the appearance of "elitism" among those who understand it. English speakers can access international data bases easily; native language speakers hardly at all. And many respondents objected, the report said, "that data of any significance would probably have political or commercial sensitivity as well, and that it would be illegal or at least undesirable to put such data into foreign computers even for one's own purposes. Through painful experience, these countries have become suspicious of developed country motivations."

The respondents offered remedies, but they affected the supply side only. Find or set up local information retailers (they suggested the Technology Resource Center in Manila). Adapt teletype networks, or lease dial lines to network nodes in other countries. Explore alternatives to private lines (as the Philippines did with a five-channel UHF system). Repatriate native experts who have been educated abroad. And, since "bureaucratic difficulties will require patience, curry patronage."

Satellites enter the picture because they treat the earth under their footprints as an undifferentiated surface. Spot beams can be concentrated into areas wholly within a country only if the country itself is able to afford such a luxury. Remember the trade-offs: as the size of the footprint shrinks, the volume of its data carriage can rise, but the number of potential users dwindles, and each pays proportionally more for the service. Currently, only the U.S., the U.S.S.R., and Canada exercise that option; China, India, Japan, Brazil, and Indonesia have some domestic satellite circuits either in place or in progress. In the rest of the world, satellites recognize no national borders, and any uplinked information can be received by any downlink antenna in the footprint. Encryption and other techniques can affect its interpretation, but the crucial fact is that, no matter what steps are taken to prevent reception, *it will be received.*

Chapter 12

FINANCING SATELLITES

"Technologically," says Glen Pafumi, "the satellite business is probably one of the safest to be in. But the cost of entry is high." Pafumi is vice president of the securities research division of Merrill, Lynch, Pierce, Fenner, & Smith. "The investment is $100 million before it produces even the first dollar of revenue. Right now, that's divided up this way: $40 million to build a satellite, $25-35 million to launch it, and $10-15 million to insure it. That range of $75-100 million is not attractive to venture-capital people: they rarely invest more than $10 million."

The satellite business is very capital-intensive, so it is almost always financed with the general faith and credit of well-established carriers. Companies such as AT&T and RCA, or governments such as those of Canada or the U.S., can afford to invest directly in satellites, to have them designed, built, and launched more or less at will, because they know in advance that there are customers ready to use or lease time on the new transponders.

"That won't change," Pafumi insists. "The first communications satellites, such as Telstar, were financed by the Bell System out of everybody's telephone bills. Since then there has been no direct public financing of satellites."

"When you look at an investment," Pafumi says, "you have to ask, 'Would a prudent man make that investment in light of the risks and the potential return?' I think he'd say it was prudent to invest in COMSAT, since it was a government monopoly." COMSAT has the exclusive right to pass American traffic through the INTELSAT network.

Merrill Lynch applied that "Prudent Man" rule in 1965 when it helped underwrite COMSAT's first public stock offering. Pafumi says that application of the "Prudent Man" rule tends to inhibit institutional investment in small businesses, particularly those run by "two guys in a garage."

Some garage-born, two-man companies have been fabulously successful (Hewlett-Packard, for example, which inspired Apple Computer's co-founders), but most cannot meet Pafumi's test. It was the small, mostly local venture capitalists (none as big as Merrill Lynch) who provided most of the growth money to Silicon Valley's computer outfits until those companies went public.

RISKS

There *are* risks in the satellite business. Chief among them, according to Pafumi, is that "a satellite begins to die as soon as it's launched. Every day you don't have a customer leasing time on a transponder represents money you can never recover."

Second, and a growing risk, is the expansion of supply. The most lucrative use of a transponder is for voice traffic, in the same way that the most lucrative advertising sales in a newspaper are its classified ads: dividing up the available space into the smallest possible units, despite the additional overhead in sales and layout stuff, produces the highest return per page. Although current technology divides a transponder into some 2000 voice channels, there are always incentives to divide it further. RCA, for example, has invested in new "single-sideband" equipment, which can fit 7200 voice channels into a transponder, though it also requires new ground equipment to do it. Under development elsewhere are "time-assigned speech interpolation" devices which can "flip" between conversations; they will carry only those in which someone is actually talking, and hold off those in which there are short pauses. If developed to practicality, that technology may double further the number of voice channels available. "Is the demand there?" asks Pafumi. "For AT&T, probably; for others, such capacity may saturate the market."

The third risk also involves trade-offs. New technology that improves the efficiency of ground stations can make some older technologies a better investment for some applications. A cable television operator who wants to receive an additional "feed" of tv programming may find it cheaper to buy a second dish than to invest in a new receiver capable of delivering two television signals simultaneously through a single dish. Says Pafumi, "When you talk about financing satellites, you always come back to supply and demand."

Criteria for investment are similar to criteria for end-use. "It's a combination of technology and the bottom line," he says.

> Is "state-of-the-art" always the best way to do something? Sometimes the old way is more cost-effective. Look at the teletypewriter for data transmission: it's slow, but it's cheap and ubiquitous, and very reliable.
>
> Do you want to be on the *leading* edge of technology or the *bleeding* edge? . . . A stand-alone word processor is on the leading edge, but trying to put a bunch of them into a network so they share common files—right now—is the bleeding edge. Maybe it'll work easier when the network control circuits have been put on chips, or when the circuit boards themselves come down in price to about ten dollars apiece. But until then, it's not worth doing. The files are stored on floppy disks, which you can take out of one machine and put into another at practically no cost.

NEW STRATEGIES

There are new ways to finance technology. One is to invest in research and development. "An enterprise can raise money by selling its R & D as a tax shelter," Pafumi says, "but that has not yet been done in the satellite area. In computer companies, like Gene Amdahl's Trilogy Corp., the investor gets a writeoff; and if the R & D pays off, he gets a royalty that's taxable as ordinary income. Since he can only write off up to the amount that's been invested, he should divide the investment among several R & D ventures to get back some of what he loses to taxes."

The cost of developing direct broadcast satellite (DBS) systems is staggering: Pafumi believes that Satellite Television Corp. spent $684 million in its first year, 1982, with no launches scheduled until 1984. Some $225 million of that came from its parent COMSAT. Pafumi estimates it will cost STC "close to a billion dollars to set up their nationwide system. That would have to be raised through bank loans, or by going public—an offering of stocks or bonds. COMSAT could raise the equity by selling securities, or selling debt, but it cannot make STC go with 100 percent debt-financing; it *could* go with 75 percent."

RCA and Western Union have also filed applications with the FCC to create DBS systems, and they are large enough to raise money through their own faith and credit, or they could float more common stock. New groups within large companies generally get credit from their own company's method of financing, provided that they can pay current interest rates.

"Satellite Systems Engineering and Direct Broadcast Satellite Corp. got a million dollars from Kansas City Southern, a venture capital company. They want to raise $30 million in equity. They hope to get $300 million from the finance-credit arm of the satellite manufacturer itself—GE or Ford," according to Pafumi. That's comparable to buying a car from a dealer who also makes the loan.

"Advanced Business Communications Corp. filed for a DBS license to serve cable tv headends and master-antennas on apartment buildings," he says. "They want to sell *options* or letters of intent to get loans from investment bankers. Hubbard Broadcasting Corp. wants to sell *participation* in its programming to independent affiliates. American Satellite Corp., which retained Merrill Lynch as an advisor, wants to build a hybrid satellite that will serve both the C and K bands and hopes to go public to do it. Unless its constituent companies (Fairchild Industries and Continental Telecom) raise their own capital, American may be the first satellite company to go public.

"Remember, it's the space segment that's expensive. Ground stations are not that big an investment. If you're a cable company, for instance, you can walk into your local bank for that kind of money," says Pafumi.

COMPETITION

Competition and technology, rather than regulation, will drive the marketplace in the 1980s, according to Jonathan Miller, editor of the newsletter *Satellite Week.* "More domestic and regional satellites are scheduled to be launched now than ever before," he says. "That will drive prices closer to cost, and will mean that the companies who launch them will have to provide their customers with something more than just the transponder. International competition is coming because America is exporting competitiveness, and the existing providers are trying to diversify. It hasn't been easy. Satellite Business Systems, which is owned and therefore funded by COMSAT, IBM, and Aetna Life & Casualty, lost $100 million a year for its first two years, and by 1983 had still not shown a profit. COMSAT is calling STC a 'strategic diversification,' but tv broadcasting is an area about which COMSAT really knows nothing. They have to hide behind their *de jure* monopoly as the INTELSAT participant, since they're certainly not the monopoly carrier for international traffic *de facto.*"

INSURANCE

Satellites are expensive. If one is lost—as happened to RCA late in 1979—it's comforting to know that someone will insure them. Brian Hughes is an executive with Inspace, an insurance company that specializes in providing coverage to the space industries. Hughes says,

> Inspace has several different types of coverage depending on what the owner or developer needs. There is *preignition* coverage, which the manufacturer typically assumes, covering construction, storage, transit, and assembly at the launch site.
>
> Then there is *launch insurance:* it lasts for 180 days after ignition. RCA received US$77 million for the loss of Satcom III because it disappeared within six months. Launch insurance covers replacement of the spacecraft, the price of a new launch, and costs incurred as a result of the delay.
>
> *Life* insurance is all-risk coverage for either partial or total failure of the satellite, starting from the 181st day and continuing for three years.
>
> There are also *liability* policies, covering third-party claims; *transmission* policies that cover interruptions or losses of transponders; and *ground support* policies to cover earth stations, dishes, headquarters, and so on.

The RCA settlement was the largest ever made; in 1977 the European Space Agency had received US$28 million, and a year later its Japanese counterpart received $12 million. According to Hughes, a "typical" 24-transponder satellite cost represents an investment of over $70 million: $35 million for the spacecraft itself, $30 million for a conventional (Delta rocket) launch, and $7 million for additional expenses.

He feels that is a reasonable amount for the insurance companies to risk on a single policyholder. But satellites are not likely to remain in the ownership of a single party for much longer. Transponders have already been sold in fee simple to corporations and organizations, as the next chapter explains, and when they all start competing for insurance coverage, Hughes says, there may not be enough to go around.

He gives this example: suppose there are 24 transponders, and it costs the operator $3 million apiece to build and launch them—that's their replacement cost. The operator keeps eight in reserve against failures by the others, and sells 16 transponders for $15 million apiece. Now, the new owners need $240 million in insurance, and the operator needs $24 million, so that satellite has to be insured—not for $70 million but—for $264 million. "A developer cannot buy insurance on behalf of the potential buyers," Hughes says. "Each will have his own requirements." The number of insurance companies willing to risk money on satellites is sure to rise, but where they will find "pools" of money for it is not clear.

Chapter 13

TRANSPONDERS FOR SALE: REAL PROPERTY IN OUTER SPACE

Until 1982, the only way an end-user could acquire time on a transponder was to rent or lease it—directly from the operator, or indirectly through a broker or reseller. Then the FCC announced that it would permit the sale of transponders outright and the industry divided into two factions. It is a rift that evokes, disturbingly, the gap between those who can afford to buy their own homes and those who cannot.

Operators were given a choice between being common carriers (providing service on a tariffed basis) or noncommon carriers (with sales or other innovative arrangements). The FCC awarded orbit slots to three companies for birds which were ready for launch in 1982 and 1983. RCA, Western Union, and Hughes were authorized to sell transponders on those birds, but future sales were not automatic, and orbit slots for proposed satellites were not assigned. The FCC reserved the right to give further approvals on a case-by-case basis.

The idea was not new. During the Nixon administration, Hughes Communications, Inc. proposed to build a private satellite system, but the idea was shelved. The FCC felt that there was a shortage of transponders and that the needs of the greatest number of users would be best met if the operators were "common carriers," in the same way that the term was applied to telephone and telegraph (and transportation) companies.

The 1982 ruling relied on the FCC's new conviction that there was no longer a scarcity of transponders. The commissioners further suggested that, for many customers, sales would be more desirable than leases. Their news release that day (July 29) said plainly:

"Transponder sales may make it possible for small users, who might be unable to afford access to transponders under a straight monthly lease arrangement, to finance a transponder purchase as a result of the tax benefits that accrue, and the ability to collateralize the transponder as a capital asset."

Generally, it has been the large or institutional customers who have welcomed the opportunity to fix their transponder costs, and their access to space, over the whole life of the bird (8–10 years).

Nonprofit groups, small, and occasional users have felt that sales of transponders would tend to raise the price of those which were left available for lease. Their attitude is based on a feeling that transponder sales will be like sales of urban real estate where, demonstrably, conversion of rental property to sale in fee simple tends to diminish the number of rental units available, particularly for low-income tenants. On the other hand, as everyone who owns his or her own house can testify, the present tax laws favor ownership because practically everything except the principal payment is tax-deductible, including interest on the mortgage and local property taxes. On commercial structures, deductible depreciation and an investment tax credit (a reduction of one's *taxes*) also encourage outright sales. Owners of even very small houses or businesses benefit greatly from the American government's sympathy toward the ownership of real property.

Indeed, the analogy with real estate is so close that Jonathan Miller, editor of the newsletter *Satellite Week,* coined the expression "Space Condo" to describe a satellite whose transponders were sold in fee, and Hughes Communications advertises its GALAXY satellite as a "shopping center," in hopes of attracting "tenants."

THE FLEXIBLE COMMISSION

The FCC ruled that transponder sales were in the public interest. The majority opinion held that, since the sellers would provide no communication service, or transmission capacity, they were not "common carriers" in the conventional sense, and therefore not subject to the FCC's regulatory jurisdiction. That position had been advanced by Hughes Communications, Inc., in its position papers.

Opposing parties argued that the 1934 Communications Act was a mandate to "provide adequate facilities at reasonable charges," and could be fulfilled only by requiring domestic satellite services to be offered to the public on a common carrier basis."

The Commission decided on what it called a "*flexible regulatory policy* [that] would stimulate the efficient and economic development of domestic satellite technology and *allow the applicants, not the commission* to shape the direction of domestic satellite operations." [Emphasis added.]

"The Communications Act was adopted long before the advent of communications satellites," the majority wrote,

> and therefore it nowhere mandates that domestic satellite operators be regulated as common carriers. . . . [W]e endeavor to ensure that the communications needs of as many diverse users as possible can be met. That many users are interested in obtaining satellite communications pursuant to noncommon carrier arrangements is evident by the number of agreements [between operators and their customers] and the pleadings before us. Thus, a decision against these arrangements would thwart the expressed needs of many consumers and satellite operators alike. Moreover, our policy of relying upon marketplace forces to shape the evolution of satellite telecommunications has proved very fruitful over the years.
>
> In any case, we believe that the shortage [which] the opponents of transponder sales aver to is a temporary one, which is now in its last stages. The industry is responding to the upsurge in demand, and it appears that ample capacity is under construction now to provide for the needs of all users, big or small. The number of equivalent 36 MHz transponders [36 MHz = 1 tv channel] will approximately double from the current 264 to 480 by year end 1984. Moreover, applications are pending for eight satellites at 6/4 GHz and 14 at 14/12 GHz. These systems could provide an additional 480 transponders, thus quadrupling available transponder supply over the next five years.
>
> There appears to be no substance to the charges that if we allow the plans for sales to go forward, it will drastically curtail the availability of transponders left for common carrier use.

Dissent came from Commissioner Joseph Fogarty who, alone, insisted that the majority's "references to a 'temporary shortage' are wholly without any rational support." Fogarty said that their perception was based "exclusively on a 'supply' analysis, which lumps together current authorized and operational transponder capacity with authorized but not yet operational capacity, as well as potential future capacity represented by pending applications, and does not in any way correlate such supply variables with *any* analysis of current or future demand. In deciding whether or not the transponder shortage identified [previously] has not been alleviated, elementary economics would suggest that *both* sides of the 'supply/demand tension' equation should be addressed. However, this decision's economic theory appears infinitely elastic." [Emphasis in original.]

Quoting from a 1974 court case, Fogarty challenged the majority's assertion that the operators could serve as regulators. He said,

> The whole theory of licensing and regulation by government agencies is based on the belief that competition cannot be trusted to do the job of regulation in that particular industry [e.g., telecommunications] which competition does in other sectors of the economy.
>
> The proponents of private transponder sales contend—and [the majority's] decision embraces the contention as demonstrated fact—that permitting such sales now, in the present climate of domestic satellite facility and service shortages, will help avoid future supply/demand imbalances by stimulating greater efficiency and capacity through more certain financing, risk-sharing, advance customer commitments, and long-range launch planning. The reality of this theory of "long-run" public interest is that the proponents of private transponder sales—and this compliant Commission decision—have only demonstrated how the applicants' private business interests and those of their select, closed group of purchasers will be served by sales approvals, not that such sales will serve the public interest as defined by the purposes and mandate of the Communications Act.

The majority opinion prevailed. The implications of its decision can be understood best by listening to the responses of those who claim to gain or lose by it.

TRANSPONDERS IN FEE SIMPLE

Since 1982, many satellite companies have financed their birds through the sale of transponders, and that has worked to the benefit not only of the sellers but the buyers, as well. Glen Pafumi explains it this way: "Western Union says, in effect, to Hughes Communications: 'Build me a 24-transponder satellite.' Then it sells the transponders, one or two at a time, to Dow Jones, Citicorp, and so on. Those companies, for example, paid $9 million apiece for transponders on Westar V. American Satellite Corp. bought 20 percent of the transponders on Westars II and III, and paid $32 million to Western Union for 20 percent equity in Westars I and IV.

"C-band transponders cost about $3 million apiece to build, launch, and insure, and I estimate that they are now selling for $10-12 million."

There have been some unusual transactions, though. "Vitalink is a company that makes redundant, nonstop uplinks; that is, they have more than one of everything electronic, so if one fails, the signal still gets through. Vitalink bought two transponders from Western Union with the help of an insurance company. But Western Union has since bought a minority interest in Vitalink, so—in effect—Western Union still owns part of the transponders that they sold!"

Though the numbers are large in absolute terms, Pafumi insists: "At $10-12 million per transponder, a company with only that much money to invest *can* get in—it's a realistic amount for venture capital firms to raise."

Pafumi expects that the current demand for transponders may fall off and be succeeded by an oversupply of transponders available for leasing. "That's probably why Western Union is selling transponders instead of leasing them: they want to protect themselves against loss from lower lease rates or from not being able to lease them at all."

To a bank, a satellite is a like piece of real estate: a mortgagable asset on which loan collateral can be based. Jonathan Miller's expression "space condo" describes a satellite whose transponders are owned outright by their users, rather than being leased to them by the builder, operator, or carrier.

"There has been an 'astral real estate' snobbery," Miller says with a smile. " 'Park Avenue', satellites like Satcom IIIR, have carried the high-class business, such as feeding 6000 cable tv systems, while the 'slum satellites,' like Westar, have had a reputation as being broken-down old birds with interruptible transponders. From now on, though, as the price of earth stations goes down, users will be driven toward the cheaper satellites, and not care who their neighbors are."

Unfortunately, thinks Miller, the hidden costs of competition may cause a shakeout in communication systems that depend on satellites. "I think that Americans may rue the day we broke up the Bell system. Local service will suffer because, by 1985, it will not cost much more to call 5000 miles than to call 500. Distance will be irrelevant to cost, making long-distance carriers very competitive. There may, possibly, be competition for local voice service, too—from cable tv systems—with the local phone company competing against *them* for the video business.

ASK THE MAN WHO OWNS ONE (OR HIS ATTORNEY)

Attorney Joel Winnik represented Citicorp in the acquisition of two transponders from Western Union. A former staff counsel for the FCC, Winnik is now in private practice. That first-ever commercial purchase in fee simple took nearly two years, beginning in 1980.

"Why would a banker want to live in a condominium with 'media' types, like cable television programmers? The answer is that modern banking is changing, and has become a function of hi-speed data communication and information processing," says Winnik. "A bank, today, is just a store-and-forward computer, a pair of wires, and a bottle of aspirin. Point-of-sale terminals have to be connected to a host computer. Citicorp owns Carte Blanche and Diners Club, and is the largest issuer of Visa and Master Card accounts. Citicorp transfers $35 million dollars every minute. Any delay could be catastrophic."

"Why did we do it?" he asks, rhetorically. "We needed to maximize control over the transfer of money. For example, our telecommunication costs were suddenly raised when a certain company reclassified its circuits from intrastate to interstate. I won't name it, but it was a very, very large company that was later to get smaller by order of a judge."

Big as it is, Citicorp did not have the money or the technical staff to launch its very own satellite. Its need for telecommunication channels, however, coincided with a desire by Western Union to sell transponders. WU had forecast that a glut of transponders during the 1980s would force it to lower its lease rates. The contracts they drew up together designated which transponders were to be conveyed, that they would operate for the life of the satellite, and that there could be options on follow-on equipment. The transfer of title was to take place a set number of days after launch. Limits were set on the operator's freedom to move or retire the bird. Resale by Citicorp was permitted, but subject to Western Union's consent. The down payment was large, with monthly, quarterly, or annual payments due on a schedule.

Out of his experience in acting to protect Citicorp's interests, Winnik has developed guidelines for prospective transponder owners: "The buyer should require the satellite to have a minimum number of working transponders before title transfers—you don't want to invest in a nonfunctioning bird. Preserve the option to terminate the contract if the satellite is not launched or operational by a certain date. Condition the transfer on the operator's obtaining a desirable orbit slot, that is, one whose footprint covers the areas you need to serve. The FCC order approving transponder sales will probably be challenged in court, and other decisions may undermine it too," he notes. "That's unlikely, but a purchaser should get a refund if regulatory approval is not forthcoming."

Winnik is not convinced that sales in fee will lead to a shortage of transponders, nor that it will encourage operators to extract high, monopolistic prices. "Assuming that the law permits an activity, the question is whether the FCC will approve it. The benefits redress the market's imbalances. Fears of shortage are unfounded because more than 500 transponders are due on stream in the next few years—also new compression techniques for squeezing extra signals through those transponders. Sales provide the cost of development and expansion, particularly in exploiting the Ku-band. Western Union estimates that 60 percent of the replacement cost of Westars I, II, and III will come from new sales. A transponder is collateral; the purchaser gets tax benefits. But the single most important advantage is that having title ensures access at a price that will never change.

"Everyone who wants to use a transponder has gotten it, including minority programmers. The market can be a very efficient allocator of capacity. Those who are opposed to sales are more concerned with price than with access."

ON THE OTHER HAND . . .

Robert Wold is opposed to the sales. He feels there is still a shortage of transponders, and he is certainly a man who has his finger on the pulse of the industry. He was one of the very first people in the U.S. to lease "long lines" from AT&T and to resell the use of them to clients on a part-time basis. For years, Wold set up ad hoc networks for sending radio and tv broadcasts of college basketball away-games to the teams' home towns. When satellites proliferated, in 1975, he was ready—with a sophisticated organization—to lease transponders and put them to work for his clients, who may need them as little as a quarter-hour a week, or as much as 24 hours a day. The Robert Wold Company now has more satellite transponder time under lease for the broadcast industry, radio and television, than does any other company.

Wold is upset at the FCC's decision on transponder sales, which he calls "the $40 billion giveaway." "That's the amount of money we, as taxpayers, invested in space exploration to make communication satellites possible," he says.

> The public should realize those benefits in reasonable prices that truly reflect the costs. Has the FCC decided that the public has enjoyed that long enough?
>
> Owners can recover costs and be rewarded for their risk-taking, but end-users can't; that's why the concept of scarcity rents was established. Should AT&T be free to charge marketplace prices for distributing television programs nationwide? The "Chicago School of Marketing" [i.e., "supply-side" economics] says the marketplace should determine price. In effect, by this action, the commission is delegating to the carriers *its* statutory responsibility to ensure that those carriers comply with the law!

Wold is downright skeptical about the FCC's conclusions.

> I do not believe that every user who can afford to lease a transponder full time for, say, $150,000 a month, necessarily has enough taxable income or potential tax liability to justify paying a $2–3 million premium to buy one. And there are many smaller users who absolutely cannot even *lease* one full-time. They don't have the income to take advantage of a transponder's investment tax credits and depreciation.
>
> Economists overlook the fact that carriers can sell off the tax benefits independently of the operating rights to the transponder, and are certainly not giving these tax benefits to the purchaser gratis.

THE SHOPPING CENTER

Jack Johnson is director of marketing for Hughes Communications. Long a builder of satellites for other operators ("Hughes supplies 80 percent of the free world's revenue-producing satellites," says Johnson), it is one of the three companies which got the FCC's green light for transponder sales.

"Since the Nixon administration began the 'open skies policy,' 15 new satellites have been launched, and 15 more are scheduled in the next five years. There are seven existing satellite companies, but several more have applications to build birds. For us to compete successfully with the established common carriers, we had to find customers whose needs would be better met by a noncommon carrier, in a condominium or purchase arrangement. The FCC allowed RCA to sell five transponders on Satcom IV, Western Union to sell 24 each on Westar IV and V, and Hughes to sell 24 each on Galaxy I and II, for a total of 101."

> Everyone now has a choice of how to acquire transponders. I think 101 will be enough for a while. Companies using full transponders in their business may prefer to buy them, as real estate, for the investment tax credits. The purchase price cannot be increased, but tariffed lease rates can. Customers cannot be shifted to another transponder, but they can be aggregated according to need or technological advantages. Competition has been enhanced by expanding the range of choice: to buy, to enter into a long-term lease, or to accept the tariff. Sales will increase new entries in the industry, since each new provider may have the option of offering sales or lease, and so gets more capital. Those new offerings argue against monopolies. And there is an active secondary market in which leasees sell transponder leases at a premium, but that gives no re-investment to the industry.

While some $400 million was invested in 1982, Johnson admits there has been "a flood of artificial orders, and speculation in transponder sales." It's an old business: "scalpers" buy blocks of tickets to popular Broadway shows and sell them at a profit to people who could not, as a result, get tickets on their own.

Galaxy I is a "plain vanilla" C-band satellite 18 of whose transponders were sold to cable programmers. Hughes calls it a "shopping center." "A successful shopping center has 2 major anchor stores, plus a number of specialty shops." Johnson says. In that analogy, Galaxy I is a major mall indeed. Its "anchors" are HBO, which bought six transponders, and Westinghouse/Group W, which bought four. The "speciality" stores are Times-Mirror, Viacom, Ted

Turner [WTBS and Cable News Network], and SIN [Spanish-language], with two apiece, and C-SPAN [the live procedings of the U.S. House of Representatives], which has an agreement for one.

The shopping center concept was obviously attractive to its tenants. "There is a real place for each of those sales," Johnson says.

> These customers want 24-hour-a-day service, every day of the year. They're interested in uptime, not down time, and our 376 Series satellites have a ten-year design life. We sold 18 "primary" transponders, backed by six additional transponders and six power amplifiers. Reliability studies showed we have a 90 percent probability that 22 out of the 24 transponders will be working at the end of 10 years. We warranted to the 18 customers that their transponders are good for 9 years. Those companies were willing to pay for protection.

As in a shopping center (or a condominium, for that matter) tenants pay additional charges for maintenance and services. "My shopping center needs to have its windows washed, its streets swept, and its ashcans taken out," Johnson says.

> Hughes performs all the housekeeping on that satellite, and we charge so much per month per transponder; the price is the same for all transponders. Remember, we ourselves are also on that satellite—on the six unsold transponders—and we share the risk.
>
> Our strategy in the Galaxy system is to work with customers to help them achieve their business objectives. We at Hughes believe that the FCC wants to stress *flexibility* to satisfy a customer's unique requirements.

In response to Robert Wold's assertion that sales will generate excessive profits, Johnson says this: "The price is my profit *for ten years*—my total profit. I can get rid of my marketing department. I'm not selling and reselling the same piece of hardware."

THE MONEY SUPPLY

It's expensive to purchase *or* lease transponders, so investment bankers will play an increasingly important role in the transactions. Gregory Clifford has participated in sales of rail rolling stock, electrical generating equipment, and other expensive capital assets, at the investment firm of Smith, Barney, Harris Upham.

"While each project is unique," he says, "they all share common elements:

1. the presence of two parties: an operator and a purchaser;
2. a tax owner who needs to shelter income—at a value up to 35 percent of the asset's cost;
3. a collateral value that gives lenders security in repayment, and which also covers maintenance, marketing, and so on.

"Financing techniques already used by airlines and railroads—project financing and leasing—will start to come into the satellite business," he asserts. "They will be characterized by three things:

1. a large cost for assets—often beyond any one party's ability to borrow;
2. tax benefits that may be beyond the ability of those building or using the asset to use them; and
3. assets that are readily marketable, providing a significant degree of the ability to finance it.

"For example, most airplane sales are financed according to the value of the airplane, not necessarily that of the airline running the airplane."

Since the mechanics for selling big-ticket items are well known, Clifford says, the real question is: "What exactly is being sold?" There is a more limited market for transponders alone than for transponders with a ground segment and other earth-station equipment.

> Look at it from a creditor's viewpoint: if he is not paid, he wants to take back the asset quickly. This is particularly important with a transponder because it is diminishing in value whether it is in use or not. Money suppliers are concerned that contractual arrangements should allow them to replace the purchaser or operator as quickly as possible—not after a three- or four-year court case.
>
> Institutions that extend credit to the *operators* want as much freedom as possible to dedicate the satellite to other users. Purchasers' creditors want to be sure they are covered no matter what happens to the operator. The value of interruptible transponders, compared with that of protected transponders, is important to creditors and investors too, as is the resale potential.

A standard sales agreement has not yet developed. The shopping center and railroad financing models are already in use, but future sales will probably allow for lower-cost financing for larger products (so more entrants can finance jointly), and for benefits that will come to those who set up their financing early. "This will change and evolve, though," Clifford predicts. "First sales may be difficult, because the financial community will be skeptical until track records are established."

WHEN THE TAXMAN CALLS. . .

Tax attorney Carl Kaseman has also been involved with tax and asset-based financing, as a former general counsel for Conrail. But he forecasts legal wrangling over the sale of transponders, since few of the terms have ever needed definitions before.

"Yes, Bob Wold, there is a Santa Claus," he says with a smile, "but he can be picky about who gets the tax benefit. The issue is not who is the purchaser, but who is the *owner,* and also there is a question over whether a transponder is a *tangible* asset? The contract determines that.

> Purchasers need to be careful. Is the person purporting to be the owner *really* the owner? Only the owner gets the investment tax credits and the accelerated cost recovery deductions. The present value is up to 35 cents on the dollar, but it must be *tangible personal property.*
>
> The existing contracts do not warrant that the purchaser will get tax benefits, they only say "we [the operator] won't do anything inconsistent with what you tell the Internal Revenue Service—tell them what you want." State taxes and property taxes may apply. Contracts should specify who pays those.
>
> Look at the lenders' interest: there are no provisions in state laws to cover lenders' rights. Transponders are integrated into a larger piece of equipment. If a purchaser wants to finance it, a lender with real dollars wants to have a "perfected security" interest. That is, if you have this asset on your books, the lender wants your other creditors to know that he has the security rights to that asset. And this may not be tested until a user goes bankrupt or defaults.
>
> If there is a bankruptcy by the purchaser, borrower, or operator, you're going to have a free for all. Unlike the laws

> governing railroads, there are no laws for satellites covering who keeps the asset, or requirements for making up the payments. How do you turn over a transponder?
>
> There is an *expectancy* that tax benefits will be transferable through leasing arrangements, but the laws are unclear right now. Property leased to the government, for example, does not qualify for tax credits, but with the lease of a transponder to the government, can the operator claim tax credits? The IRS has said, so far, that it's a *service contract,* not an interest in the property, because the operator retains possession and control, and the transponder is preemptible — the operator can substitute another transponder to satisfy the service contract. Since the FCC did not provide for sales before, sales as such do not exist in service contracts.
>
> Is a transponder a tangible asset? Is there a concept of exclusive use over life? Is there a risk of loss? (There are refunds and partial refunds that negate a true risk of loss.) Are sales *final?* These questions remain unanswered.

Some tax benefits accrue in a lease-back arrangement. Clifford suggests that if the FCC does allow transfer of ownership, and the IRS calls it tangible personal property, and the purchaser can be classified as the owner, then a lease-back arrangement can be made. A bank, or other lending institution, can claim to be the owner and lease back the equipment to the purchaser. "What is the nature of this property interest? Is this a lease of 'limited use property'? The IRS says, where property is leased for its entire usable life, or with no prospect that the leassor can repossess and resell to others, the leassor cannot be distinguished from a lender, and the tax benefits belong to the leasee. If you can't repossess, then the resale rights of the lender—on default by the borrower—will determine whether it is 'limited use property.' Safe Harbor lease laws may cover this, if new laws allow it."

As for title, Clifford says, only the bill of sale testifies to a transfer of ownership, unlike titles to automobiles which are recorded by government agencies. When does title pass? "To claim the investment tax credit," he says,

> you have to be the first user—and the time frame is when the asset is "ready" for use, not when use actually commences. If a satellite were launched, and the transponder

were ready, but title passes later, then the IRS may say "It was ready for use before you or your leassor acquired ownership." So the satellite operator can claim the credit—but that does them no good; if you claim the credit and then sell the property, either the profit is recaptured [as tax] right away, or you're no longer eligible to claim the credit.

In short, the contract documents we've seen will quickly grow from four or five pages to 15–20 pages to deal with these issues.

IS THERE A "DOWNSIDE RISK"?

Not everyone is enthusiastic about transponder sales. Besides those resellers who stand to lose some business, such as Robert Wold, there are skeptics even among the operators. Lee Paschall is one. He is president and chief executive officer of American Satellite Company, which provides digital services to the private and public sectors; its government clients include NASA, the armed services, and the Department of Defense.

Paschall is unequivocal. "We will consider long-term leases for our transponders, but we do not believe in selling our assets. We won't sell."

Even the FCC was initially hesitant, because it was not convinced that sales would be made in a way that was fair to all potential buyers. In late 1981, RCA placed seven Satcom IV transponders on the auction block at Sotheby Parke Bernet, alongside antiques and other works of art. The winning bids ranged from $10.7 to $14.4 million each, but the FCC soon invalidated the auction sales on grounds that common carriers are not permitted to discriminate with respect to price. It is one of the original tenets of regulation, dating from canal-boat days, but FCC Chairman Mark Fowler called it "a severe government restraint on private business."

Another FCC official was later quoted in the *Wall Street Journal,* giving this suggestion for nondiscrimination: "You could invite sealed bids from 50 applicants, choose the top seven and then charge those seven the same price—the lowest bid among them." It wasn't necessary. By summer, 1982, the Commission voted to allow direct sales to whomever the operators wished.

Carriers and operators point with concern to the practice of leasees selling all or part of their transponder time. "Prime time available immediately!" shouted one ad in the *Wall Street Journal.*

"Now get in on the ground floor of the new era in satellite communications. Save money and hassle with smaller receiving dishes. Cut through terrestrial microwave interference." It seemed to have been written by the same people who make late-night-tv pitches for obscure franchise opportunities, commemorative plates, or incredibly sharp knives. But the advertiser was the American Hospital Video Network, which wanted someone to take over 108 hours per week from its lease. Between 7 p.m. and 7 a.m. Monday through Friday, and all day Saturday and Sunday, the Network apparently could not generate enough traffic on the SBS transponder for which it had contracted. How or why the Network chose to lease a transponder for 168 hours a week, which it needed for only 60, was not explained in the ad.

No one knows yet if competition will drive prices up or down. Another advertisement, which ran in the *Wall Street Journal,* the *New York Times,* and *Variety* in late 1982 offered a "Satellite Transponder For Sale": fully protected service at a discount price, with low down-payment and generous terms. The Washington, D.C., attorney who had placed the ad for his client (who remained anonymous) said he received "over two dozen" responses, "fully half of them serious." What he almost certainly was "selling" was the right to use the remainder of a long-term lease, but—as the saying goes—possession is nine-tenths of the law.

Fortune magazine ran a surprisingly downbeat story in November, 1982. Staff writer John Cooney called transponder sales "one of those grubby, cost-conscious markets that new players enter at their peril." Despite their attractiveness, he warned entrepreneurs, "Pass up the transponders. They are ugly little devils at best, and you may well be able to buy them cheaper next year."

Robert Wold still insists that prices will go up. "Those of us who got into business early, who committed to a lot of transponder time, saw a lot of lean times. We had to go out and market that time at a very low cost, just as the cable tv companies had to offer cheap, basic service in the early 1970s. Once satellites were in place, and able to bring in vastly greater programming options, the value of a cable system got hyped. Now, the same thing will happen with satellite capacity itself. Admittedly, as we get into the '80s, we'll have more K-band capacity, so prices may change.

"But remember this: I don't oppose the *idea* of selling condos, instead of leasing transponders, but I believe it's too early. Later in the 1980s, when there are more satellites and transponders, the operators may be able to go unharnessed. For now, though, we still have a scarcity situation, and a hype situation where the people with the satellites will be drawing unrealistic returns on their money. Carriers are—by tariff—normally permitted to realize a 20 percent after-tax return, under existing lease arrangements. With sales, they will realize 40–50 percent return after tax."

Wold's argument rests on the nature of the orbital arc as a resource: he sees it as ultimately *public* and not private. "The operators were given very precious orbital slots. At the time, they said they would operate under regulation, as common carriers, providing reasonable, cost-based communication services. That's what they said, to get those slots, and we believe they should live up to the promises they made."

Section 5

NEW OPPORTUNITIES

"In my view, it is possible that we will be able to call distant cities anytime, at low cost, but we'll be wait-listed for a local dial tone!"

—Jonathan Miller
Editor, Satellite Week

Chapter 14

DIRECT-TO-HOME BROADCAST SATELLITES (DBS)

Thousands of people in North America are already getting their television programs directly from satellites. With an unobstructed southern view toward the geostationary arc, all they need is the money—about what they'd pay for a small new car—to buy a 3-meter dish antenna, a "low-noise amplifier," and a "downconverter." The dish collects the satellite's signals the way a rain gauge collects water; the machines with the exotic names process those signals and turn them into the stuff that their tv sets can display.

It's become a thriving business for savvy electronics sales-and-service people throughout the U.S. and Canada, especially in regions too far from the cities to receive regular tv broadcasts. Weekly and monthly publications list the programs that the C-band satellites will be carrying, and are packed with advertisements for more dishes, amplifiers, and converter boxes.

The dishes and their electronics come as kits (the famous Heathkit® makes one), or as "turnkey" systems (like a new car, you just turn it on). For as little as $3000, a family can watch what the rest of the world watches: soccer matches and foreign-language news. Or they can see for free full-length motion pictures that cable-tv subscribers have to pay to watch. Some people enjoy Johnny Carson's talk show untouched by censors: live, as it happens, from Los Angeles, while it is being sent by satellite to be edited in New York.

Home satellite dishes are part of a video revolution as powerful as the computer revolution. Videotape recorders, particularly video cassette recorders (VCRs), have drawn audiences away from commercial network television. By purchasing or renting prerecorded tapes, and by recording tv shows for viewing at a more convenient

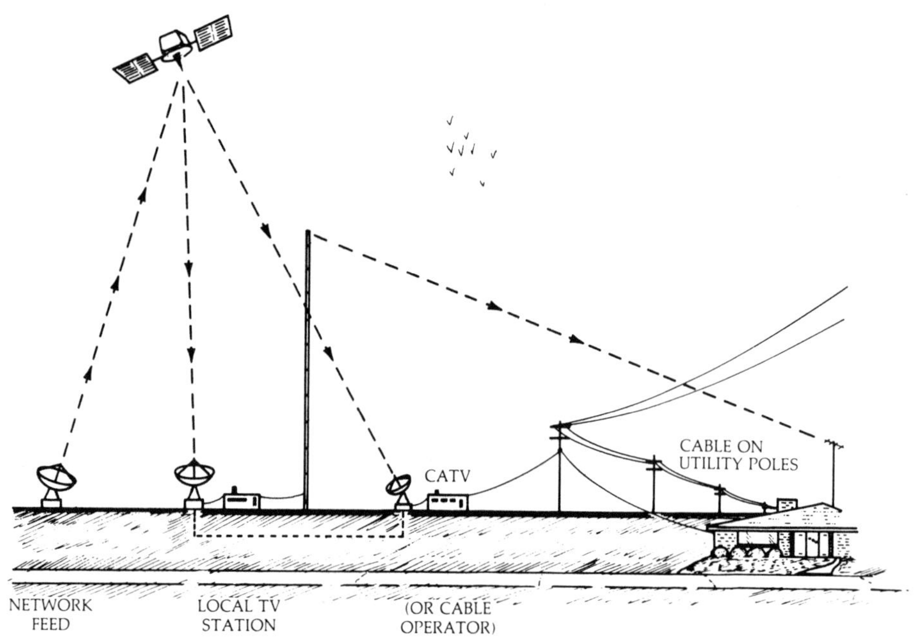

The easiest (and still the most common) way to receive television pictures by satellite is through local tv stations or cable companies' facilities.

time, VCR owners have turned television from a supplier's medium into a consumer's medium. Sales of video disk players have not been as spectacular, probably because they do not make recordings, but the machines and disks are cheaper than VCRs and tapes are; and too, they make movies and special programs available *at the viewer's discretion,* rather than at the broadcaster's whim. Distributing tv programs by cable has made it possible for a viewer to choose between a dozen, two dozen, or even three dozen or more channels. For the first time ever, in 1982, broadcast network executives admitted to the FCC that they had lost viewers—not just in percent but in actual numbers.

WHY ISN'T IT HERE ALREADY?

The next wave of the video revolution is direct-to-home broadcast satellite (DBS) service, but before people ride that wave, the hardware will have to be much easier to live with than the present "generation" of home-satellite receivers are.

Most people can't or won't put a C-band dish in their backyards. A 3-meter dish, after all, is as big as a Volkswagen beetle. It has to be anchored in a concrete foundation, aligned with a compass, and periodically adjusted. Few suburban householders have that much room, even fewer have the southern exposure as well, and practically no city-dweller could ever put such a machine on the roof. Yet it is the urban and suburban populations whom advertisers and broadcasters want most to reach.

Cost is also a factor. Unless a dish is under the satellite (i.e., near the same meridian of longitude, and not too far north either), the signal falling to earth is very, very weak. It is diluted with "noise:" spurious radio signals from local microwaves and from outer space itself. Isolating the signal requires a very sensitive electronic circuit called a *low-noise amplifier* (LNA) that is both delicate and expensive. The farther away from the satellite a dish is, the more sensitive (and costly) the LNA must be. The LNA *can* be less sophisticated but, then, the dish will have to be proportionally larger, so it can collect more signals for processing—an inexorable trade-off that bangs its head against the argument for smaller dishes.

Only the downconverter box with the channel-selector knob on it is a familiar piece of hardware, very similar to the converter box that cable tv systems use. But wherever cable tv is available, the hookup is made by a professional installer, for a very small fee, and the householder gets a low, monthly bill; it's a far cry from the complexity of rigging one's own antenna. So the principal customers for home satellite antennas—so far—have been rural people with big yards and no other tv sources, plus the usual gadget-freaks and the very rich who will buy one of everything.

True DBS services will be broadcast over a much higher frequency band than the current satellites use, and will therefore accept smaller dishes. A mass-market for those dishes is imminent, but it waits for three developments, all of which are due by 1985: mass-produced, cheap dishes; high-powered satellites; and competitive programming.

DISHES

If DBS satellites are to compete with local tv and with cable, they will have to have antennas that are *at least* as easy to install as a tv antenna or a cable hookup. Potential DBS providers, in their applica-

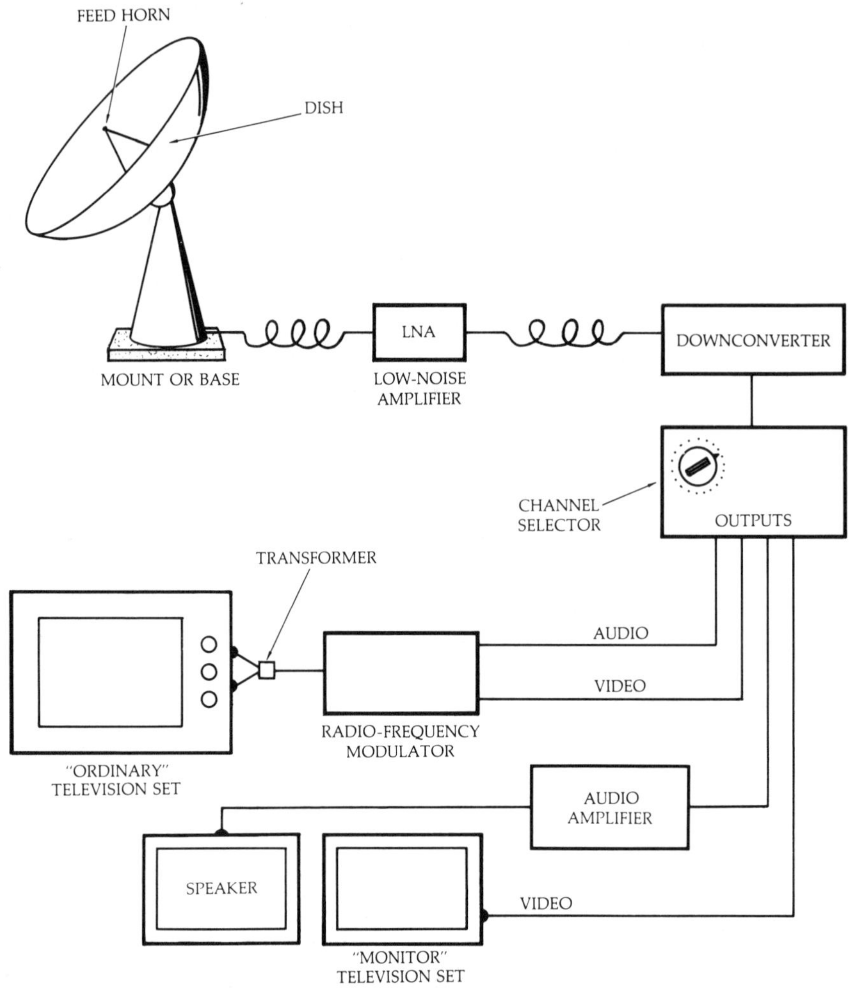

A simplified satellite tv system.

tions to the FCC, have all advocated small dishes, a meter or smaller in diameter, that can be stuck on a rooftop, lashed to a chimney, or tied to a pole in the yard. DBS antennas can be compared with rooftop tv antennas, but one difference is the criticality of alignment. A householder knows more or less where the local tv broadcast tower is, relative to that house, and can point the home antenna in its general direction. The signal is so powerful, anyway, that practically any grounded metal stick will pass along enough of it to make an acceptable picture on the tv set. A poorly aligned tv antenna can be rotated by one person while another checks the quality of reception. DBS

Log cabin with a view of the world: *In rural areas far from tv stations and tv cables, DBS may fill the video needs of North American families. The small antenna is possible because the satellite is very powerful and uses very high frequencies. If Abe Lincoln had had one, he might not have had to walk 12 miles to school. (Courtesy COMSAT)*

providers hope they can make the dishes that easy to use, but satellites will be much harder to find than nearby towers, and adjustments to the LNA and downconverters may be much more critical than those of their counterparts in an ordinary tv set.

Mass production of dishes will be easy. The technology has been around since the 1940s, and has been improved by new plastics and plastic-handling machines. Some manufacturers claim to have already "struck" a mold for DBS 1-meter antennas. But, until firm orders come in, no one wants to make them on speculation, and the price will depend on volume. At least at first, someone—probably the DBS carrier—will have to subsidize their manufacture and installation.

RAINY DAY BLUES

DBS systems will not use C-band; they will use the 17/12 GHz band called Ku-band (pronounced "k-u") which is much higher. That's a technical advantage, since it is less susceptible to terrestrial interference and background space noise, but it can be blocked by heavy rain. In order to concentrate the power of its transmissions, a

When DBS services begin, antenna owners will have to align their dishes much more exactly than they do with their regular, terrestrial tv antennas. Professional installation is a potential growth industry. (Courtesy COMSAT)

transponder using the Ku-band must cover a much smaller footprint than a C-band transponder can. Therefore, more than one satellite will be needed to cover the whole of the continental U.S. The westernmost of those satellites will have to have special "spot beams" for Alaska and Hawaii since, although they are sparsely populated compared to the mainland, they are supposed to be served. The FCC requires interstate communication tariffs to be nondiscriminatory with respect to the states: that is, Alaska and Hawaii must have the same services at the same prices, despite the fact that serving them is more costly to the carrier.

THE SHADES OF NIGHT

During the year, especially in the summer months, there will come a time every day (roughly around midnight, local time) when the satellite is in the earth's shadow. These "eclipses" are as short as a few minutes and as long as half an hour. To continue relaying programs, the satellite has to draw on its batteries for power. But having reliable batteries that are capable of supplying uninterrupted power to the transmitters with the same strength as the regular, solar power cells, requires the bird to be heavy and costly. If the operator chooses to let the satellite go "dark" during the eclipse, then the viewer may lose the DBS signal around 11 or 11:30 p.m. That's still "prime time" in many places, and if it happens too often, viewers will become disgruntled. The operator may be able to supply the signal via backup satellites, but that too is costly and may require the receiving dish to be temporarily realigned—hardly what the householder wants to do around midnight.

THE GOOD-NEIGHBOR POLICY

Another consideration is the number of DBS satellites, and their positions in the geostationary arc. That depends very strongly on the outcome of a conference beginning in mid-1983, in which all the nations of the western hemisphere will meet in Geneva under the auspices of the ITU to allocate DBS orbit slots and frequencies. The U.S. hopes that the Regional Administrative Radio Conference (RARC) will give it permission to launch satellites into four primary and four back-up slots. Each slot could support eight or ten Ku-band satellites without interference, with the potential for delivering a total of 30 to 40 different tv channels of programming.

The four primary slots correspond to the four time zones across the U.S. Canada, being larger, wants five. Mexico and all the other Latin American countries, except for Brazil, are each longitudinally narrow enough to be served by a single time-zone satellite. Notwithstanding that, they are "developing" countries and—almost without exception—want DBS allocations made *by country* so that they will be guaranteed a slot, whether they can use it right away or not.

The U.S., of course, wants allocation to be made—not on a one-for-one basis but—according to how many slots and how much frequency spectrum a country can be expected to use. Allocation *by use* would allow the U.S. (and Canada) to launch several satellites at once, giving the equipment manufacturers a tremendous headstart on the market, which will surely grow throughout the hemisphere in the 1980s. Indigenous industries in other countries are helping their governments to lobby for allocation *by country,* so they can get a share of the business. It is very, very sticky. The chances are, though, that the North Americans will find some way to go ahead with their plans, more or less, no matter what the outcome of RARC in Geneva.

THE WASTELAND . . . AGAIN

The last hurdle that DBS must overcome is a lack of programming. The dilemma is similar to the one RCA faced shortly after World War II. It had the factories to make television sets and owned the studios from which to broadcast programs, but neither business was paying its own way. If RCA tried to sell tv's, people said there weren't enough programs; if they tried to get sponsors for programs, the advertisers said there weren't enough people with sets to watch. David Sarnoff, in a brilliant stroke, had free sets installed in bars and taverns, and directed his programmers to televise baseball games. Within a year, RCA was selling both sets and airtime briskly.

Cable operators thought they could get away with showing sitcom reruns and other castoff shows from the broadcast networks, but found their greatest money-makers were recent motion pictures and otherwise-unavailable sports. The creative and performing arts have had occasional moments of success on cable, but they more often languish in limbo—with cultural, educational, and scientific programs—because intellectuals are embarrassed to admit that they watch tv, and program directors do not believe that the masses are "ready" for art.

What, then, will DBS programmers offer? The Direct Broadcast Satellite Corporation (DBSC) is one of four to which the FCC has given permission to go ahead and make a system. According to its filings, and its promotional literature, it is making deals with consumer groups such as the National Citizens' Committee for Broadcasting; educators, such as the National Education Association; organized labor, including the United Auto Workers and the International Association of Machinists (both of which, by the way, have substantial membership among aerospace workers).

DBSC is also looking toward Heaven. "The National Catholic Telecommunications Network proposes to use the DBSC system to interconnect all the Catholic dioceses in the country, enabling them to share programming and services. Other religious organizations that have expressed an interest in the DBSC system include the

United Church of Christ and B'nai B'rith International," according to DBSC.

Satellite Television Corporation (STC), a division of COMSAT, says it will carry "a diverse selection of popular entertainment, sports, children's, educational, cultural, public affairs, and minority-oriented programming." That description could apply to what the major networks and the "superstations" (such as Ted Turner's famous WTBS, Atlanta) already carry.

Indeed, there is some evidence that DBS may simply enhance the network's dominance over programming. If the networks' signals go by satellite directly into people's homes, which is entirely possible, they will not have to go through a local "affiliate" station. That means no preemption by local programmers, no substitution of local advertisements for national ones, and no hassles from local citizens complaining about sex, violence, and crass commercialism. The viewer will see pretty much what he or she already sees on the networks. Local stations, of course, are complaining about the potential loss of "local origination" and "community involvement" which they profess to offer their viewers now. Yet neither they nor local cable operators have ever been able to make local programming pay, except for local news and local sports. It is no secret that many radio and tv stations would rid themselves of public-service announcements, public affairs programs, and other things of purely local interest if their federal licenses did not require it.

The FCC has, on several occasions, come close to reinterpreting that vestige of canal-boat days, the requirement to serve "the public interest, convenience, and necessity," in such a way as to relieve tv broadcasters of responsibility for presenting programs of community concern. In the name of "free speech," the FCC and the Congress may one day eschew completely the notion that the communication channels ought to serve the public interest.

Several commissioners have said they believe that if people are really interested in their communities, then local programs, made by amateurs with inexpensive equipment, will compete for the viewers' attention with network programs employing professional talent and state-of-the-art machines. It has never happened before; but they seem to feel that with the proliferation of cable-borne channels, minority-owned, low-power tv stations, and videotape alternatives, community programming can keep itself afloat without federal regulations. For the time being, however, local programming is still a condition of license.

HOW MUCH IS THIS GOING TO COST ME?

If DBS carried network "feeds," it would have to let subscribers watch them for free; advertisers would pay for the programs and, by extension, for DBS itself. If *subscribers* have to pay, then evidence accumulated by cable tv operators suggests that viewers will not pay for programs which are interrupted by commercials. Most people will accept ads between shows, but not inside them. The innovative two-way cable QUBE system, in Columbus, Ohio, has had modest success with "infomercials," which are five-minute presentations similar to those in a "magazine-format" tv talk show or a do-it-yourself demonstration; they employ or display the sponsor's product and end with some kind of pitch. A related idea, in many cities, is a cablecast video tour of houses for sale, each of which is tagged with a real estate agent's phone number. DBS could certainly become a common carrier for those kinds of advertiser-supported programs.

But no matter what the programming, the subscriber will have to pay *something* to get DBS. STC proposes to charge $100 for an antenna, including installation, and about $14.00 to $18.00 a month (in 1981 dollars, which is when they filed with the FCC) for the converter box and other electronics. Outright purchase of the equipment will also be offered, and the price will be "several hundred dollars." United Satellite Communication says it will use a Canadian satellite (ANIK C-2), over a transponder leased by General Telephone and Electronics (GTE), to carry movies, sports, and news. Basic service would cost about $15.00 a month, plus the lease of the antenna for about the same price; buying it would cost the subscriber about $600–700 (in 1982 dollars).

There were more than a dozen applications to the FCC, asking for a license to provide DBS service; most of them were from small or medium-size companies hoping for a chance to strut their stuff. Two of them—this really happened!—were from convicts in prison. Four organizations eventually got the FCC's nod, but some controversy still swirls around one of them. Satellite Television Corporation, being a subsidiary of COMSAT, is viewed in somewhat the same light that wholly owned telephone companies used to be, with respect to the Bell System. Competitors worry that the huge parent will take revenues from its most profitable enterprises (in COMSAT's case, the INTELSAT traffic) and use them to subsidize the fledgling until it is solvent. That cross-subsidization would be a competitive advantage for STC, but it could also weaken the financial health of COMSAT if it proved to be a drain. STC and COMSAT deny vehemently that it would happen, noting that STC has its own $400 million line of credit with a group of some of the largest banks in the U.S. But the FCC had to protect the interests of COMSAT customers—after all, it is through COMSAT and through COMSAT alone that Americans send telephone, telegraph, facsimile, and computer data via INTELSAT to the rest of the world. The FCC ruled, in its approval order, that COMSAT may not give any additional funds to STC, nor extend credit or assume debts beyond those that the original filing declared, without further authorization by the FCC.

Commissioner Anne Jones, the only dissenter from that 1982 ruling, was quoted at the time as saying that the other applicants would provide comparable service without "the tremendous risk to rate payers" that COMSAT might incur if it had to bail out STC. The FCC's Common Carrier Bureau produced a memorandum on the subject also, which questioned two of STC's assumptions about the DBS business: that venture capital would be available at a 10-percent rate of interest, and that the service would have 650,000 subscribers at the end of its first year in operation. After years of watching cable operators revise their subscriber projections downward, few observers are inclined to be that optimistic.

AS IF THAT WEREN'T ENOUGH . . .

The chief competitors of DBS service may not be cable systems, but subscription-tv and over-the-air pay-tv which are distributed in many cities by microwave. These services are usually marketed to apartment houses and other clustered dwellings with master-antenna tv (MATV) systems, in which one antenna supplies signals to the whole residential complex. Single-family houses may also subscribe, for about the same price as what DBS proposes to charge, though the economies of scale favor MATV installations. Customers get movies, mostly, from HBO, Showtime, Cinemax and The Movie Channel; occasionally, they are offered one-time entertainment or sports events.

The systems are more or less automated, and the technology has been around for 40 years, but theft is a serious problem with pay-tv that will certainly plague DBS. Since the signals are transmitted over terrestrial microwave frequencies, and there is a healthy supply of new and surplus microwave equipment on the market (with no FCC restrictions on using it *for receiving*) innovative merchants have packaged all the necessary equipment into convenient kits. By buying the equipment, customers need not pay the program supplier anything, and lawsuits are being filed every day.

At issue, of course, is *who owns the signal?* So-called "pirates" say that once the electrons wiggle up and down in the antenna and make waves through the air they are in the public domain, just as music broadcast through a radio can be heard by anyone who chooses to listen. System owners, naturally, claim that pay-tv is *not* like radio, but like telephone or telegraph traffic: *private* communication between a sender and one or more receivers, and that it is protected from unauthorized interception by the Communications Act of 1934 and other statutes. Canadian trials (see Chapter 4) have set no legal precedents there, and even if U.S. courts eventually rule that pay-tv *is* private, there is very little the FCC or anyone else can do to stop piracy. Local police could arrest people who sell, buy, build, or install the microwave kits, but few prosecutors would ever be able to prove that the defendants intended to steal pay-tv signals with them. It is clearly impossible to stop people from trying to beat the system.

The alternative—for subscription tv and for DBS—is to scramble the signal, to encode it, somehow, so that only authorized, paying customers will have boxes to decode the program. That is very expensive, unless the signal is all-digital to begin with; digital encryption is relatively cheap. But the cost of making analog tv into digital signals, transmitting, receiving, and converting them back to analog form (since analog is what ordinary television sets require) would almost certainly be prohibitive. Also, the reality for pay-tv companies is that, wherever they have tried scrambling their signals, whiz-kid pirates have built de-scrambling machines and sold them to other pirates in open defiance. The code-makers and code-breakers of the world are locked into their little war, just as the armor plate and projectile people are.

DBS *is* coming, but the ordinary citizen who only wants a wider choice of tv programs may have to wait a while for the dust to settle.

Chapter 15

PICTURES FROM SPACE

Satellites are not *only* for communication in the telephone and television sense of the word. They are also used as cameras.

Some picture-taking satellites ride in the geostationary orbit, but most of them are in "low-earth orbit," on the fringe of the atmosphere. That's where the first astronauts' Mercury project capsules went, and where the Space Shuttle spends most of its time. Spacelab, the "space station" that fell out of the sky in 1980, was in low-earth orbit.

Gravity decreases smoothly and gradually the farther an object moves from the surface of the earth, as Newton realized. But there is still some atmosphere above 20 miles, which slows down the movement of objects. Satellites whose orbits take them lower than 100 miles have lifespans measured in days or weeks. By comparison, geostationary satellites have "lived" for 18 years and longer.

Low earth orbit is the home for most of the picture-taking satellites. Some follow the equator, others zip around the poles or follow bizarre but vital orbits so that they can be placed over strategic places at specified times. The military potential for low-earth orbit is the subject of great controversy and is beyond the scope of this book. Suffice it to say, lasers and other antisatellite weapons, if they are ever used, will be used first in low-earth orbit.

A NEST OF HAWKS

The first balloons that could carry man-size loads were pressed into service during the Napoleonic wars. Floating above musket range, their passengers were spies who noted the enemies' positions and strengths, and then dropped report packets to the generals below. It was 100 years before the civilized Europeans used airships to drop bombs.

In 1960, an American pilot was convicted of espionage against the Soviet Union after his "spy-plane" was shot down. The "U-2 incident" drew attention to overflights, and to the fact that photographic science had so improved that usable and strategically vital pictures could be made at a distance of more than a dozen miles.

A 5-ton Soviet satellite with an atomic power plant scared a lot of people in 1978. Cosmos 954 had been routinely surveying ocean traffic from 150 miles up, in a polar orbit; and just as routinely, when its useful life had ended, it was being boosted up to about 600 miles—a parking place where its radioisotopes could fizzle out their half-lives in no particular hurry. But something went wrong, and the Soviets unhappily had to alert the world that 110 pounds of fissionable uranium, complete with strontium, cesium, and iodine byproducts, were going to crash, burn, and fall out someplace on earth. Fortunately for almost everybody, Cosmos 954 burned up over Canada's Northwest Territories and scattered its poisonous cargo into the January snows.

Four years later, a nearly identical scenario unfolded, but Cosmos 1402 fell in the ocean. World attention was drawn to the atomic-powered satellites, and the Soviets fiercely denied that there was any danger. Indeed, as they publicly predicted, the reactor and its containment shell caused no damage as they fell, but a wave of emergency-preparedness swept Europe and Canada.

In order to fly so low, picture-taking satellites have to fly very fast; also, half of every orbit traverses nighttime darkness. So they cannot spread wings of photovoltaic cells to catch solar energy and turn it into electricity, as communication satellites do. The Viking Mars lander, and the Pioneer and Voyager spacecraft which hurtled past Jupiter and Saturn, also cannot draw power from the sun because they go too far away. They all carry thermonuclear batteries—actually tiny generating plants—which stay very hot inside as a result of the heat of fission, and very cold outside because of the vacuum of space. The temperature difference can be turned into electricity by the action of special materials in what is called a *thermocouple.* The fissionable materials selected are those whose rates of decay match the satellite's expected lifespan.

A former CIA officer was arrested in 1978 for selling the Soviets a technical manual for the operation of "Big Bird," one of 12 camera satellites that had spent five years looking over Russian missile silos, airfields, and troop maneuvers. The KH-11 satellite took electronically enhanced pictures on special film and ejected the exposed capsules over friendly territory, on parachutes, at its lowest orbit altitude—about 90 miles. CIA sources noted, at the time, that with "Big Bird" photography they could distinguish between civilians and persons in military uniforms, call out the make of an automobile and read its license plate. The Americans also acknowledged that year that—like the Soviets—they, too, had lost a couple of nuclear-powered surveillance satellites. One, they said, burned up over the Indian Ocean in 1964 and scattered its fallout "worldwide."

ANTISATELLITE WEAPONS

The conventional wisdom, on both sides of the Iron Curtain, is that before any military threat can be carried out, the opponent's early warning systems must be rendered inoperative. Since each country looks down from space on the other, it seems likely that the real "first strike" weapons will be those which blind the other's eyes.

Satellites are curiously vulnerable to attack. Those with solar power arrays need only to lose their orientation to lose power; those with thermonuclear batteries have delicate antennas, camera lenses, and other sensors that can be knocked out of alignment. A satellite does not have to be entirely destroyed in order to be rendered useless.

Though their orbits are far out in space, the technology for reaching those orbits is very well established; hundreds of satellites, large and small, are in orbit already, with dozens more added each year. What is important is the fact that those orbits, once established, are rarely changed; they are predictable, and easily simulated by large computers. But one of the hardest ways to knock out a satellite would be to launch something else into its orbit in such a way that the two will eventually collide. By the time such an object were really able to inflict damage, the operators of the satellite would have detected it, tracked it, and made adjustments to move their satellite out of its way.

A more subtle weapon would be a simple explosive device, launched into an orbit *near* the target satellite. If it were detonated close by, say, within a few miles, fragments would probably damage the satellite's sensors, communication, or telemetry antennas. Television documentaries produced in the late 1970s, which drew on U.S.

government files, claimed that the U.S.S.R. has tested such a weapon in space against a dummy target of its own,

At about the same time, the Soviets publicly expressed concern that American Shuttle astronauts would "kidnap" foreign satellites, holding them in the Shuttle's cargo bay either for examination, tampering, or return to earth. Since the Shuttle operates only in low-earth orbit, such an activity would probably be confined to picture-taking satellites, if it ever occurred at all. A "trojan horse" scenario, with vulnerable satellites booby-trapped to explode in the hold, is a possible response to such a threat.

The trouble with explosive or physically damaging antisatellite weapons is that they must be in fairly close proximity to the target. If laser weapons were used, instead, it would not be necessary to maneuver them any closer than a few hundred miles. As every hunter knows, a moving target must be "lead" by the bullet: one aims ahead of the target's position, in the direction in which it is traveling, so that the bullet reaches the target not where it "is" but where it *will be*. A laser beam travels at the speed of light, so one merely gets the target in sight and presses the trigger; there is no appreciable delay between the time the laser leaves the "gun" and the time it reaches the object.

Lasers can damage a satellite in several ways: mainly, they can focus intense heat onto sensitive mechanisms that are not ordinarily shielded. Alternatively, they can inflict physical damage by boring through the skin of a satellite to burst tanks of stationkeeping fuel. Laser weapons are as likely to be used in geostationary orbit as in low-earth orbit, but would probably have to be tested first in low-earth orbit where the effects could be monitored by geostationary tracking and relay system satellites looking "down" on the proving ground.

Laser weapons have been tested on earth, but the atmosphere distorts or reduces the power of such focused light waves over relatively short distances. Even "laser cannons" mounted on tanks will be ineffective if the battles in which they are engaged produce any smoke, dust, or explosions. In space, however, the range of a laser beam is theoretically limited only by the line of sight and by the power of the laser device itself. Not all lasers require electricity; some chemical reactions produce a comparable lasing of light; those chemical lasers, by the way, are more likely to be used in space than electrical ones, because their constituent ingredients weigh less than the equivalent electrical power pack.

THE EAGLE EYE

Some of the so-called "spy satellites" drop their film back to earth for processing, but there is another technique for taking pictures at high altitudes: they can be transmitted and processed electronically.

All-electronic sensors are like television cameras. They can record digital data that represents each "pixel" (the smallest picture element, like the "grain" in a photograph). The satellite transmits that data to dishes on earth as a series of numbers, one pixel after another. Computers examine each pixel and—according to the programs under which they are running—assign a value to it. That value can be replaced with a certain intensity of light on a video screen or a photographic facsimile—say, the higher the number the brighter the light. A whole black-and-white picture (or, more accurately, a "grey-scale" representation) can be assembled from individual dots.

Color photography is done in a roundabout way: three separate "scans" of the same scene are made—one through a red filter, one through a yellow filter, and one through a blue filter. After they are processed as three gray-scale pictures, they are printed with colored inks—just as newspaper or magazine photos are reproduced—to generate a full-color photograph.

In practice, though, they are not made true to the "real" colors of the scene, as a human eye would see it. "False color" is used instead, to enhance features that are not apparent to the eye normally. In a civilization long accustomed to reading maps where adjacent countries are painted with contrasting colors, it is surprising how many people find false-color images disturbing. Those from LANDSAT are the most eye-catching; almost everyone has seen them: agricultural and forest lands come out red; cities and other paved surfaces come out blue. That is because LANDSAT images are processed to enhance infrared light from earth. Growing plants reflect and also give off infrared light; paving stones absorb it. Variations in the intensity of the infrared from vegetation is an indicator of its distribution, its density and its photosynthetic efficiency. Forest managers can see at a glance if timberland is dying. Cement and other building materials absorb infrared, so their LANDSAT image is "cooler" and thus darker.

Other picture-taking apparatus can include radar, a special kind that looks at the earth "sideways." So-called "side-looking" radar can penetrate sand and topsoil to reveal ancient contours. Archeologists have mapped long-abandoned settlements in the upper Nile desert by following the dark streaks of old roads that radar images have uncovered. Seen in both visible and infrared light, certain geological features also are only apparent from very high up. Prospectors compare LANDSAT maps of possible mining or drilling sites with those of existing mineral deposits and oil fields; similarities can suggest new places to dig.

LANDSAT rides in a low polar orbit which—because the earth also rotates—puts it over the same place on the ground every 18 days or so. Since it is possible that at least one of those overflights will be at night, two LANDSATS are used, in comparable orbits, but nine days apart. Thus, observers can count on seeing the same place at least once every 18 days, if not every nine. This makes possible monitoring over time. The changing boundary between land and water shows up especially well. Each year, during the monsoon season, the governments of India and Bangladesh map the extent of flooding in their great river deltas. The U.S. Geological Survey updates its contour maps with LANDSAT now. The spread of cities into the suburbs, or the disappearance of oases from the deserts, can also be tracked this way.

Hawaii, for example, uses LANDSAT data for coastal zone management. Instead of receiving color pictures, the state uses a computer to generate special maps. Pixels of deep water, shallow water, and water overlying mud or coral are printed with unique symbols; new maps are compared with old ones, and any changes due to

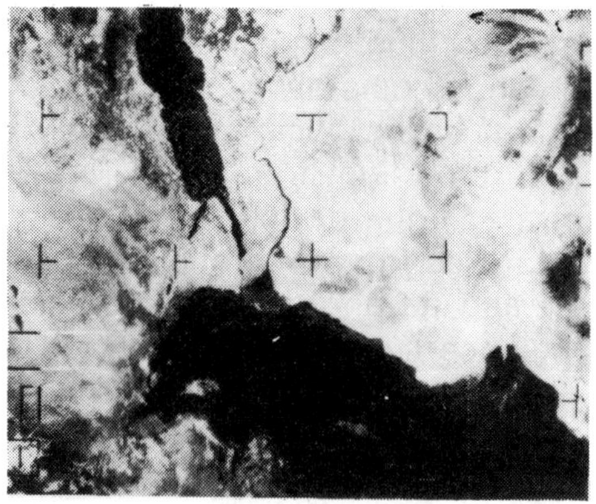

The Middle East—*as seen by a satellite and transmitted electronically to earth.*

tides, currents, and human activity such as dredging or pollution can be seen immediately.

(Incidentally, Hawaii receives those LANDSAT images indirectly. The original data is downlinked to the LANDSAT imaging headquarters on the Mainland; the data that Hawaii needs is isolated there, packaged, and retransmitted over an old NASA communications satellite [ATS-3]. It is received in Hawaii on VHF radio [149.22 MHz], using fairly simple antennas similar to those used by PEACESAT conferees on ATS-1.)

The space segment cost of LANDSAT picture-taking is more or less fixed, but the ground segment cost is rising fast. Administration and overhead, especially, may eventually cost more than the U.S. government is willing to spend on them. The current thinking at NASA is for the agency to get out of the LANDSAT business and turn it over to the private sector: in NASA's words, to "commercialize" the operation. Negotiations are already underway with the private sector to assume the cost of Atlas/Centaur and Delta rocket launch vehicles, with NASA becoming, in essence, a general contractor for launches. Similarly, operation of LANDSAT—and first crack at its pictures—may appeal to mineral-extraction and other resource technology companies. But touchy questions remain about who actually *owns* those pictures, and under what circumstances the private organization would have to share them with the public or with their competitors.

A report from the congressional Office of Technology Assessment doubts that space picture-taking has enough commercial potential to survive without government subsidies, yet COMSAT has already asked to buy LANDSAT's operations: its satellites in the sky, one that isn't launched yet, and all four of the government's weather satellites. COMSAT says that the U.S. government is the principal customer, since it already uses 95 percent of the data from space. COMSAT proposes to combine all the remote sensing and imaging into what it calls the "Earthstar" system.

Other companies want LANDSAT, too; the Commerce Department's National Oceanic and Atmospheric Administration is evaluating their proposals and may have a decision by the mid-1980s. Ironically, while getting the government out of the imaging business appeals to the Reagan administration, the idea actually originated in the Carter administration which preceded it. NASA's own plans are to continue to launch LANDSAT birds, with increasingly sophisticated imaging sensor "eyes" for near-infrared light as well as visible light, that are expected to work through about 1988.

Other countries, too, want LANDSAT-type imaging. Some, like Japan and Brazil, may launch their own remote-sensing satellites in this decade, in hopes of attracting revenues from domestic resource-extraction and management sectors. Other countries, like Bangladesh, have people whose lives literally depend on advance warning of natural disasters. It dovetails with Third World goals to make developing countries *participants* in space—not just customers for it.

A SURE SIGN OF RAIN

As with communication, there are trade-offs in picture-taking. LANDSAT is so high up that each pixel represents 1.1 acres. A LANDSAT photograph covers an area 115 miles on a side. By contrast, airborne cameras on a U-2 can examine just a few square miles, but in greater detail. The application determines which mode to use. Government agencies which operate over large regions, land-use planners, and resource managers prefer LANDSAT's broader pictures; those who are concerned with smaller regions, and those who need to zero in on trouble spots that have been located by LANDSAT, use aerial reconnaissance.

Geostationary orbit is not ordinarily used for picture-taking, because the resolution is low. With a footprint fully one-third of the earth's surface, details are necessarily lost. But the "big picture" can be of supreme importance if the phenomenon under examination is also as wide as an ocean—the weather, for instance.

On Thanksgiving day, 1982, Americans in the western states were piqued to learn that the principal weather satellite over the Pacific had gone dead. The deceased GOES-4 took a series of infrared pictures all day and all night which could be displayed sequentially, like an animated cartoon. These showed the actual 24-hour movement of cloud systems, fronts, and so forth, as they drifted eastward. GOES-4 pictures were widely broadcast on local television news and weather reports, often with computerized false-color imaging, for clarity, and gave quite a bit of accuracy to West Coast forecasts.

An on-board electrical failure apparently caused GOES-4 to stop just as it was retracing an image for transmission to the ground. Fortunately, the failure came just days *after* a hurricane had damaged the Hawaiian islands. Fortunately, too, an earlier weather satellite (GOES-1) was still hanging over the Pacific. It was seven years old, and could no longer take pictures at night, but it still worked by day. The National Oceanographic and Atmospheric Administration (NOAA), which operates the GOES birds, was not able to launch the replacement GOES-6 until April, 1983. Hughes Aircraft, which built the $37 million GOES bird, was apparently responsible for the electrical failure, and so did not receive from NOAA an incentive payment of some $3.5 million which it would have gotten if the satellite had lived out its full life.

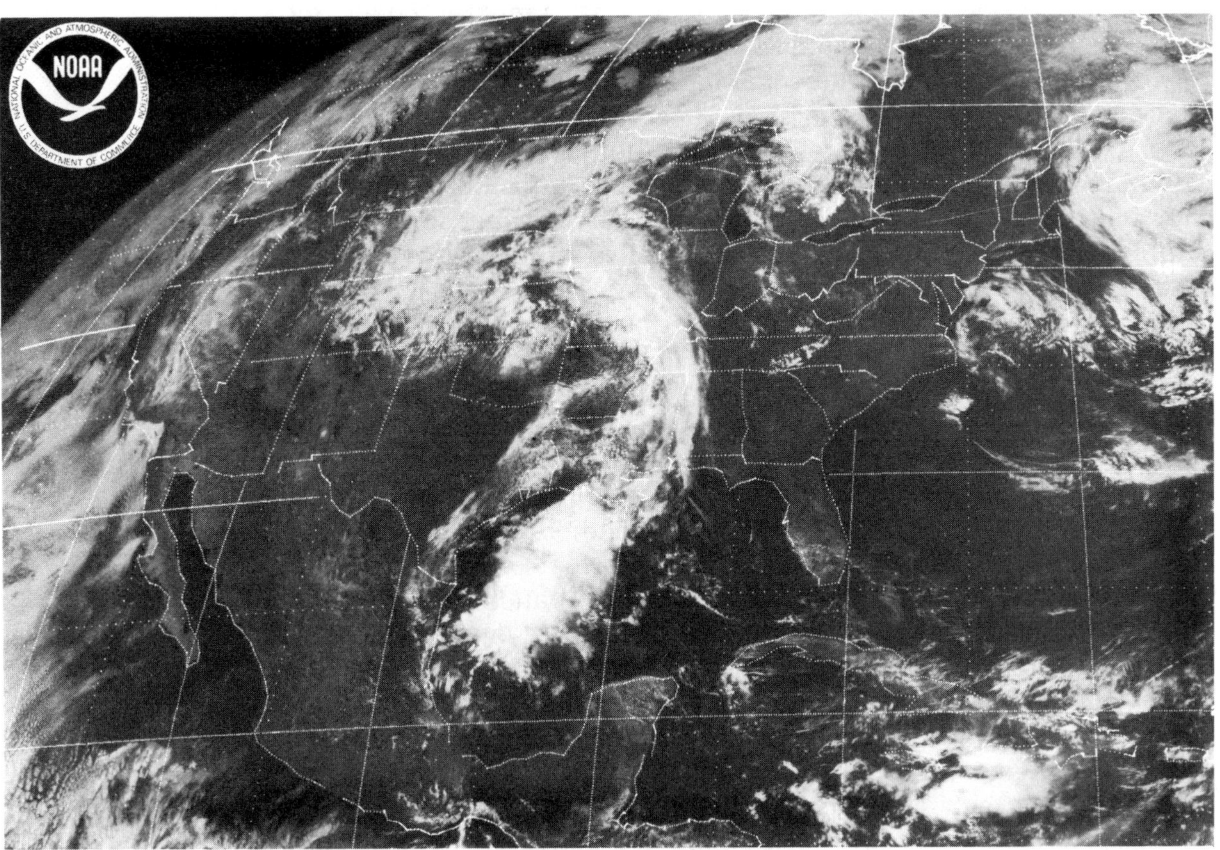

This is the GOES EAST daily weather picture for May 14, 1982. Layered frontal cloudiness stretches from the western Gulf of Mexico to the lower and mid Mississippi Valley, and then to the northern and central Plains states, producing rain showers and thundershowers. Broken low-level clouds are off the southern New England and mid-Atlantic coasts, and over eastern Maine. Thunderstorms cover the western Gulf of Mexico. High and middle-level clouds cover Nevada and portions of California. Mostly clear skies are over most of the East and Southwestern U.S. (Courtesy NOAA)

Some polar-orbiting satellites were also called upon to fill in for GOES-4, by taking pictures over the north Pacific shipping lanes. GOES-5, which is over the Atlantic, still works all right, so weather systems that reach North America can still be tracked across the continent. But GOES-5 can only see as far as the West Coast, on the fringe of its footprint, and the people who want to know what storm systems are moving in from over the Pacific will just have to wait for GOES-6 and keep their own "weather eye" out.

GOES satellites do not take their pictures at random intervals; the images are very carefully timed to occur at precisely the same time of day, every day, for good sequential data. That requires them to have clocks on board that are extremely accurate, and those clocks, in turn, permit GOES birds to fulfill another useful function: they transmit the exact time to anyone who cares to listen. An "atomic clock" in Boulder, Colorado, is the official timekeeper for the U.S., expressing the rate of decay of a radioactive isotope as a series of radio pulses every few millionths of a second. The on-board clocks of GOES satellites are synchronized with the Boulder clock, and are regulated in orbit in case they get too fast or slow. The U.S. has broadcast an official timekeeping pulse via shortwave radio since the 1920s, and the GOES satellites expand the range of that service over both oceans.

The U.S. armed services use GOES satellites, as well as their own, for meteorology. "Three-to-five-day weather forecasts are often the deciding factors in scheduling missions that require cloudless, unperturbed weather," says Lt. Col. Henry W. Brandli, USAF (Ret.).

During the Vietnam war, says Brandli, an installation at Tan Son Nhut Air Force Base required four large antennas and receiving hardware that filled two semitrailer truck vans. The "transportable" version of that equipment weighed 32,000 pounds, required two electrical generators, and had to be ferried by giant cargo aircraft; it took a team of four men 11 hours to assemble it.

The Air Force subsequently developed a genuinely transportable earth station for weather data. According to Brandli, it weighs 25,000 pounds, requires only one generator, and is fitted with wheels so it can be hauled by rail or truck. Six men can assemble it in four hours. Like its big brothers, it generates reports from computer data that can be displayed as facsimile or video images. "Ground conditions can be interpreted from enhanced satellite imagery, which indicate such phenomena as flooding, snow cover, sandstorms, and other terrain information. The importance of such data is underscored by the botched hostage-rescue mission to Iran (in 1980), which was hampered by a sandstorm. Nighttime, low-light visual imagery reveals additional important factors, such as illuminated clouds and smoke from ground fires, which indicate wind flow and city lights. Intervening inclement weather can also hamper the use of sophisticated electronic weapons such as tv bombs (missiles guided remotely by video) and laser devices." Says Brandli, transportable earth stations have made military meteorology "come of age."

The very fastest computers in the world are at work predicting the weather. Starting with GOES and other data, the so-called "supercomputers" (made by Cray and Control Data Corporation) experiment with thousands of possible outcomes from known facts.

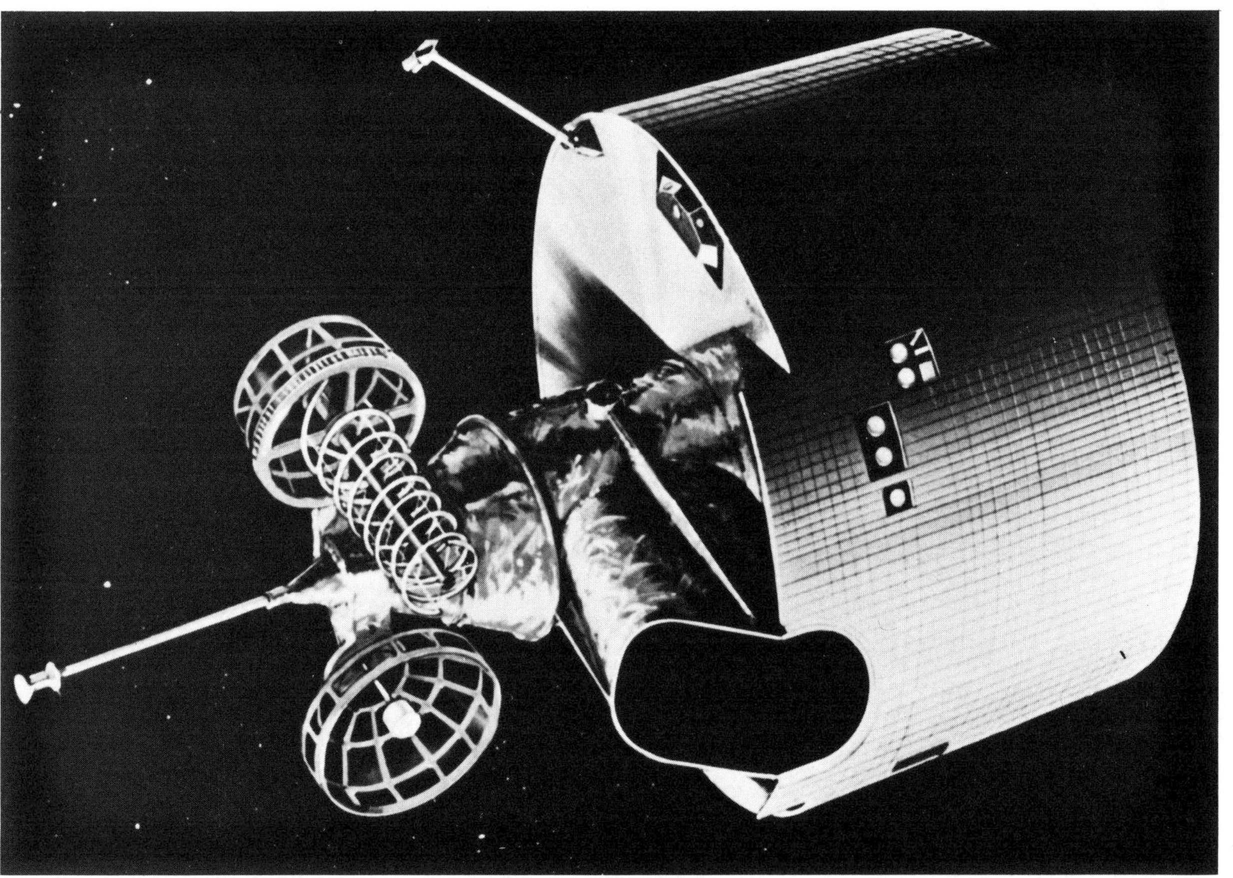

The Weather Eye—*The GOES satellite sends photographs of Earth's weather to the National Oceanic and Atmospheric Administration. (Courtesy NOAA)*

The programmers make hypotheses about phenomena, plug in the GOES reports, and direct the computers to project them forward, to see if the actual weather resembles the computer's projections. The success rate is improving, but the bottom line is still that the weather is unpredictable.

Chapter 16

POSITION-LOCATING SATELLITES

Sailors say that what's been called "dead reckoning" is really a colloquialism for "de'ed reckoning"—a contraction of "determined reckoning"—which is finding where you are by applying the procedures of navigation. The science is ancient.

WHERE IN THE WORLD?

Pacific islanders sang mnemonic poems to identify currents and trade winds; that—plus astronomy—gave them the skill to make predictable landfalls on almost every habitable rock and atoll. The magnetic compass (a Chinese invention) freed Mediterranean ships from having to hug rocky shores; the first reliable chronometers helped eighteenth century explorers map the world.

In the twentieth century, radio was employed as a method of navigation. If a ship received a signal from three distant points, it could obtain a compass heading for each, and triangulate its position from the intersection of the lines. LORAN, as it is called, is of greatest value far out to sea, where no landmarks, lighthouses, or other reference points can be found. Closer to shore, it is harder to be in range of more than one or two LORAN stations at a time, and extreme proximity to one transmitter can distort signals from others.

The U.S. armed services has, since the late 1970s, developed and tested a satellite system that helps a person to determine almost instantly his or her exact location practically anywhere on earth. Though the design is classified, the theory is similar to LORAN. A series of low-earth orbit satellites emits radio pulses that are actually timing signals, coordinated by extremely accurate atomic clocks aboard the satellites. A soldier on the ground wears a backpack that is both a receiver and a computer; when it receives a signal, it calculates that satellite's track and stores it. When it receives a signal from a second, and then a third satellite, it triangulates the soldier's position from the intersection of the "lines" it has plotted.

The system is alleged to be so accurate that the soldier's position can be accurately stated *within ten meters in any direction, including up and down.* In other words, with the appropriate hardware, the system could be used for aeronautical navigation, too. Military satellite systems, such as Navstar, can be used to guide missiles to selected targets, both by monitoring their trajectories and by feeding that information back to the missile's on-board computer for navigation. However, the armed services may be having problems with sophisticated systems like those; in 1982 they announced that a civilian version of the receiver/backpack could be made available, but it would have an accuracy only within 1000 yards. By degrading its accuracy, they hoped, the true accuracy of the system would remain classified. But LORAN is at least as accurate—sometimes more so—and is cheaper and more widely available.

HOMING PIGEONS

True navigation by satellite would be expensive, and would require fairly sophisticated equipment; a more affordable technology is now used for locating foundering ships and downed aircraft. Since the early 1970s, ships and airplanes have carried miniature radio transmitters that will send out a signal in case of emergency. One which is triggered by the impact of a plane crash is called an emergency locator transmitter (ELT) and one which is triggered by immersion in water is called an emergency position-indicating radio beacon (EPIRB); each can also be set off manually.

The principal patent for the ELT is held by Stephen Glatzer [the author's cousin]. A private pilot and electronics inventor, Glatzer developed the "beacon" idea into a business in the 1960s. In 1970, the U.S. Congress passed a law requiring general aviation aircraft (small, private planes) to carry ELTs, and since 1972, the International Civil Aviation Organization has required that aircraft making long flights over water monitor the EPIRB frequencies. The beacon is a battery-powered transmitter in a rugged box about the size of a cigarette pack; the signal is triggered automatically by a "gravity switch" inside it, sensitive to the g-force of a crash landing. Its frequency,

121.5 MHz, is monitored worldwide by aircraft flying long routes over water, and by search-and-rescue operations practically everywhere. The armed services use ELTs and EPIRBs that broadcast at 243 MHz. International agreement on emergency frequencies is one of the great successes of the ITU.

If a ship sinks, or a plane goes down, rescuers can home in on the device, which will emit its signal continuously for two days or more. However, ELTs and EPIRBs have one serious limitation: VHF radio propagation is limited to line-of-sight and cannot pass beyond the horizon. A rescue ship or plane has to be within about 300 km to pick up the transmission; rough terrain, especially valleys, can severely limit the range of an ELT.

SARSAT, the Search-And-Rescue SATellite, is an attempt to overcome that obstacle. The U.S., Canada, France, and the U.S.S.R. have collaborated to build a series of six satellites which, from 150 miles up, monitors the ground or sea below for 121.5 and 243 MHz signals and relays coordinates of those signals to a ground station any time that it finds them. Of the six satellites, only the Soviet satellite has been placed in orbit.

SARSAT works on the principle of the "doppler shift" which is familiar to most people when they hear a train whistle: the pitch seems to rise as the train approaches, then it falls as the train moves away. As a satellite approaches the vicinity of a downed airplane, it will "notice" that the ELT signal frequency is slightly below 121.5, but rises as the satellite gets nearer. The signal will be exactly 121.5 when the bird is right overhead, then it will seem to fall below 121.5 as the satellite moves away. Interpreted by a computer on earth, that would provide ground-tracking stations with a line, east to west, along which the transmitter will be found. The low-earth orbit satellite takes approximately 90 minutes to circle the earth, at which time it will take a second fix on the ELT. The earth will have rotated on its axis too, so the new doppler shift will produce a second east-west line at an angle to the first. The ELT will be found within about 25 km of the intersection of those lines. While not pin-point accurate, SARSAT will at least direct airborne search-and-rescue teams to the right vicinity, where their own monitors can home in on the accident site.

Experiments with doppler-shift position-location were made over the past few years. The Double Eagle balloon, attempting to cross the Atlantic, was ditched near Iceland in a storm; rescuers followed data from NASA's experimental Nimbus-6 satellite. Nimbus also tracked

Naomi Demura, a Japanese explorer who carried a beacon on his dogsled during a trek north from Greenland to the pole.

The first real-life use of SARSAT came in 1982, when a Canadian single-engine plane, suddenly out of power, glided into a narrow valley in British Columbia. It carried an ELT with an external antenna, but that wire had been sheared off by trees during the crash. The people aboard, who were hurt, managed to jury-rig an antenna; SARSAT found the signal first, and rescuers quickly found the people. ELTs that require external antennas are more vulnerable than those which contain a miniature antenna within their box; however, if it survives the impact, the external antenna is more efficient for transmitting and can reach greater distances.

There are other limitations to the ELT, principally, the fact that they can emit false alarms. A poorly mounted device, or one which is mishandled, can begin transmitting even though the aircraft is not in jeopardy; rough landings—those which cause slight injury—can set off the ELT even though the plane or the people aboard are in no danger. ELTs are built to be tamper-proof, so it is not easy to switch the transmitter off. SARSAT cannot distinguish false alarms from real ones any better than ground-based or airborne search teams can, and in any region with a lot of air traffic, several ELTs may accidentally go off.

To minimize the effect of false alarms, NASA proposes that new ELTs, designed to work with SARSAT, use a much higher frequency (406 MHz). Though it would take more battery power, the higher frequency could be transmitted *intermittently* (say, a burst for one second, at 100-second intervals). A false alarm would still be possible, but the signals would not conflict with one another. Suppose you are addressing an auditorium full of people. You may say, "When I count to three, everybody shout his or her name aloud"; it will be impossible to hear any individual name. But if you say, "Look at your watch, and when it says exactly 12:17, shout out your name," each person's watch will be slightly out of sync with the others, and each name will be more distinctly heard.

However, 406 MHz is not yet widely used and, in the U.S. anyway, planes are already required to carry ELTs that operate at 121.5 MHz, and which cost between US$200 and $600. Higher-frequency transmitters would be at least twice or three times more expensive, and that is a limiting factor for getting private pilots and pleasure-boat owners to accept them. However, the higher frequency could carry more information, and that could speed up a rescue. For example,

ELT and EPIRB beacons could transmit not only a homing signal but identification of the vessel (wing number or ship registration number), country of origin, user class (pleasure boat, commercial aircraft, etc.), and even a code representing the nature of the emergency, and whether medical help is required.

Widespread adoption of 406 MHz as an emergency frequency would permit SARSAT to operate with the highest level of precision: NASA claims that it could locate a distress site to within 2-5 km (1-3 mi) and, when fully operational, could detect and identify between 200 and 400 different emergency signals at one time, worldwide.

Personal protection could also come from satellites, if Gerard K. O'Neill and Charles Schmidt are successful in obtaining FCC approval for Geostar. These men propose to trigger public-safety and law-enforcement action against muggers and other animals by having citizens use a satellite to relay emergency signals from pocket-size pagers. O'Neill, a Princeton University physicist, and Schmidt, an RCA satellite engineer, say that the pocket unit could print a reassurance message from the sky: "Stay put—help is coming."

INMARSAT wants to run a similar geopositioning system on a grand scale, by which a ship would transmit a unique identification code continuously. INMARSAT's geostationary satellites would receive and interpret the signals, plotting the ship's movement and relaying those positions back to the ship periodically. If the ship were lost, or if it were foundering, its location would be known immediately.

Conceivably, systems like these will be available to the general public late in this decade, or in the 1990s. The on-board processor will have shrunk in size and cost by then, as will all microelectronic circuits; the demand may or may not materialize, but the idea will certainly appeal to independent travelers.

Chapter 17

DON'T COUNT CABLES OUT

One of the world's leading authorities on submarine cables says, "When I see a new satellite launched, I rejoice—there's a scarce resource being used up!"

SATELLITES ARE NOT THE ONLY ANSWER

Bogumil Dawidziuk is manager of market development and product planning for Standard Telephones and Cables, Ltd., of Great Britain. To him, any prediction that communication satellites will replace the need for cables is simply not based on the facts.

"Many services *prefer* cables," he says. "From a security standpoint, signals are enclosed in a structure that does not radiate or broadcast electromagnetic energy; cable-borne messages cannot be picked up by unauthorized people, as satellite transmissions can."

Cables also represent the shortest path that data can travel between two points. Relaying a signal by satellite means a near-50,000-mile round trip: even at the speed of light, that takes a quarter of a second—enough to require special equipment at both ends to ensure synchronization and accuracy.

So don't count cables out of *any* long-distance service yet. Not while Australia, New Zealand, Fiji, Canada, and the U.S. are planning major new transatlantic and transpacific cables; not while new switching technologies are making existing cables more efficient; and not while fiber-optic cables are poking their way into the water.

THE WIRED OCEAN

The Atlantic was cabled first, because that's where the traffic was. The behemoth *Great Eastern* was the nineteenth century's first ocean-going cable ship, and the telegraph cables it laid were still working at the turn of the twentieth century. Because we, today, take long-distance telephony for granted, it may come as a surprise to learn that the very first transatlantic telephone cable was not laid until 1956; before that, short-wave radio was used.

The first Pacific telephone cable was laid in 1962, but new ones are being planned all the time. According to Martin Fournier and Nirmal Pal, of Teleglobe Canada, the ANZCAN project will connect the north and south Pacific countries. It will have begun service in late 1983, in part to replace the 1960s-vintage COMPAC cable system that is nearing the end of its design life. COMPAC's 82 circuits (3 kHz bandwidth each) have been increased by multiplexing technology; taking advantage of time zone features, it now has the equivalent of 333 voice circuits. By comparison, ANZCAN will have—from its inception—the equivalent of 480 voice circuits along its *shortest* link (New Zealand to Norfolk Island) and a total of 1380 connecting Australia with Canada; those will be 4 kHz circuits, which will make data transmission much easier.

Standard Telephones and Cables is the major contractor. "For wideband data traffic," Dawidziuk admits, "cables are expensive right now, because analog bandwidth is expensive. In the ANZCAN project, 1033 repeaters and 65 equalizers will be placed at the bottom of the Pacific Ocean. Five sets of terminal equipment will provide the necessary power supply, signal amplification, equalization, and multiplexing." It's worth it, he says, to provide 25 years of reliable and stable service.

THE LIGHT AT THE END OF THE CABLE

Dawidziuk says that although ANZCAN will be a conventional, analog coaxial cable, future cable projects will probably be all-digital, with fiber optics taking over as more providers gain experience with it.

The first deep submarine fiber-optic cable was laid in 1981, under Lake Washington, in Seattle. With 35 strands of glass, each no thicker than a human hair—and more transparent than water—the cable can carry as many as 134,000 separate voice channels. Pacific Northwest Bell paid $15.00 a foot for the 11,000-foot cable, and incurred costs of about $6.00 a foot laying it down. The lake bottom, at 200 feet, is deeper than a diver can go without special equipment, and so the Seattle cable represented the first chance for a utility to

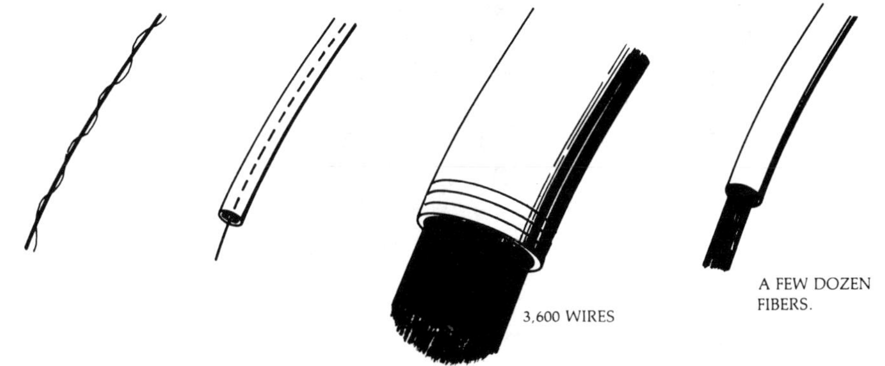

The reason why fiber-optic cables will ultimately replace copper.

practice new cable-laying skills that fiber-optic technology will demand—principally a more delicate touch at the helm of the barge and tug, to keep the glass from "snaking" or snapping.

Dawidziuk predicts that the first transoceanic submarine fiber-optic cables will be laid around 1988, between the U.S. and Europe, with "relief for the shortage of cable circuits between Hawaii and the continental U.S. following soon afterward." These new cables would carry advantages of their own, such as underwater branching points for separate landings and easy mingling of data with voice traffic. "Enhanced system economics will come from significantly increased repeater spacing." This is because glass fibers are so pure, there is less loss than with copper. Fiber-optic cables need fewer repeaters.

"The world is gradually moving from analog to digital in all forms of telecommunication network operation and hardware," he says. "Undersea systems are well established, and—despite some indecision in the early 1970s—annual investments in cable technology worldwide have averaged more than $200 million. In fact, between 1974 and 1982, $1.8 billion was invested in cable systems."

Along short but thickly-populated routes, fiber-optic cables will probably replace every other form of "trunk" line, including satellites. AT&T completed such a link between Washington, D.C. and New York in 1983, with service to industrial parts of Massachusetts and

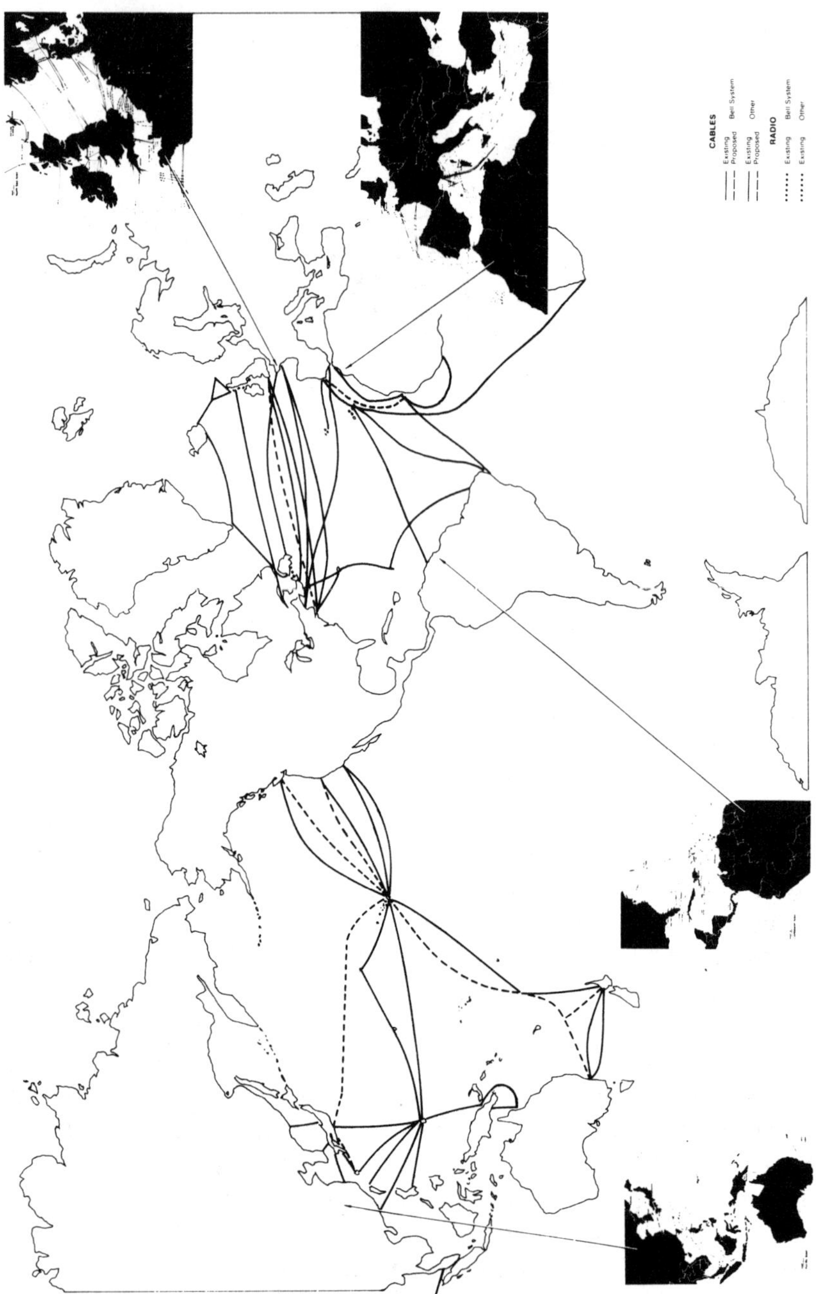

The proliferation of undersea cables—despite advances in satellite technology—is evidence that some applications are better handled by wire. (Courtesy AT&T Long Lines)

Virginia expected within the following year. Over less than 500 miles, fiber-optic cables make a lot of sense: they can carry far more traffic than a satellite transponder can, so—even though they cost a little more per circuit to build—the number of potential customers on them is greater; the carrier gets paid back sooner, and users are not as likely to have to wait for an open line.

So, even as the world looks up to the satellites, its most heavily trafficked routes, and its shortest, fastest links, are more and more likely to be cabled.

Chapter 18

WE ARE SILENT

"How would you feel if your television were suddenly taken away? The question seems odd because in this day and age we tend to take modern communication media for granted. Newspaper strikes do happen, and storms can knock out power to radio and tv stations, but they are temporary interruptions, and there are alternatives. But suppose there weren't. How would you feel?"

That challenge comes from Vijay Ram Trehan, a communications engineer and doctoral candidate from India who found out what did, in fact, happen when their satellite was moved, and television service to five million people stopped.

There have been many academic studies on the impacts of tv on communities that get it for the first time. The most recent was done by Wilbur Schramm, in Samoa, who found—not surprisingly—that people tended to watch a lot; the hut with the tv became a new community center; traditional conversation and discussion groups withered; and children demanded the products they saw on the screen.

There have also been occasional studies of tv "withdrawal symptoms" in developed countries, the most famous of which was conducted by the Knight-Ridder newspaper chain and called "The Day They Took Away TV." But it, and the others, were either of a short duration, or were made in artificial environments. In any case, in those countries, the participants knew that the experiment would not change their lives.

Trehan's research focused on people whose lives *were* changed.

India had asked the United States for the loan of NASA's ATS-6 (the sixth applied-technology satellite), which was the first satellite capable of broadcasting television programs directly to small earth stations. By being high-powered, the satellite could concentrate the television signal into comparatively small areas—spot beams with footprints about 200 miles in diameter were overlapped to cover the interior of the subcontinent. High power aloft allows low power on the ground. Antennas were made of inexpensive materials, and could tolerate less-than-critical shaping, so many of the 3-meter dishes were built in the people's own villages, by local handymen, using chicken wire. The electronics for the receiving equipment (UHF-TV) came from components supplied by the government. NASA moved ATS-6 by telemetry from a position over Brazil to one over India.

The project was called SITE, the Satellite Instructional Television Experiment, and its programming was mainly educational. News and instructional programs—and some "recreational" entertainment shows—filled the 1300 hours that were broadcast between August, 1975 and July, 1976. The news was broadcast in Hindi, while the other material came dubbed into the languages spoken by the target villages. Transmissions were made for 2½ hours each night.

As in most rural places on earth where there is television, people watch it as a community activity. The average attendance for SITE programs was about 100, after the initial novelty wore off. Some 30 percent had never seen tv before, anywhere, and—in general—more women watched than men.

The Indian Space Research Organization surveyed the villagers and found that watching educational tv improved their understanding of health and hygiene, politics, and family planning, and that those gains were proportional to the amount of tv that the people watched. Trehan interviewed villagers during a three-month field trip in 1979, three years after the SITE experiment had ended. He selected two SITE villages to study: one which had been able to replace those programs with terrestrial tv broadcasts, and one which (for geographical reasons) could not. The "tv village" was Mallepally, and the "non-tv village" was Alipur; both near the Hyderabad-Bombay highway.

Although the two villages were demographically similar, and their awareness of SITE nearly identical, Trehan noted that Alipur is near a large Christian missionary school, and that 20 percent of its people are teachers whose literacy and mass media "consumption" make Alipur's overall literacy level somewhat higher than that for Mallepally.

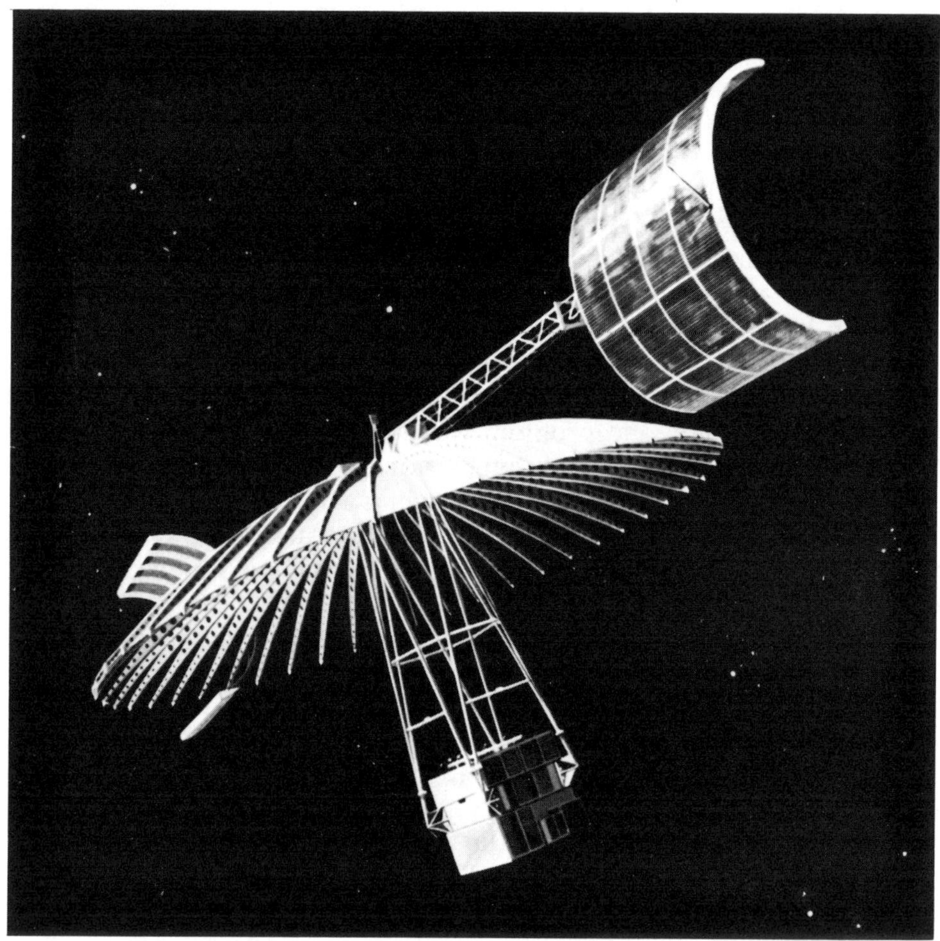

This is ATS-6, the satellite that broadcasts educational programs over India. (Courtesy NASA)

"People in the non-tv village watched primarily for education, while those in the tv-village watched mainly for agriculture [programs] and entertainment. It is interesting to note," said Trehan, "that most respondents in the non-tv village thought they had watched television longer than they actually did. That exaggeration is probably a manifestation of their loss."

A significant difference between the two groups came in their attitudes toward SITE itself. In retrospect, the non-tv villagers found no reason to dislike SITE programs, while nearly two-thirds of the tv-villagers listed complaints about programming; they had, of course, subsequent tv experience with which to make a comparison.

"The majority in both villages felt that the content and quality of the SITE programs in no way justified the withdrawal of television from their villages," Trehan found.

Associated Press reporter Paul Chutkow, writing at the time SITE ended, found that it had brought "fundamental changes in their methods of farming, cooking, health, and hygiene—and, perhaps most important, teaching" to villagers in the Rajasthan desert, in northern India. " 'I use soap now, and a cloth towel,' " a 13-year-old boy said.

P. V. Krishnamoorty, then secretary-general of Indian television, told Chutkow that, "One problem was with the school science program. Village teachers simply were not capable of explaining the basic lessons being broadcast. 'We will be increasing our teacher training,' [Krishnamoorty] added, 'and we are encouraging the teachers to use simple everyday things such as kites and windmills to explain scientific concepts.'

"The evening newscast of the government presents positive news of Prime Minister Indira Ghandi's administration and virtually no criticism and no views of opposition political groups. But Krishnamoorty said the broadcasts are in no way meant to influence the villagers politically. 'They're a test bunch. They think for themselves in any case,' he said." Chutkow reported that the government planned to continue educational television broadcasting, as much as possible, with terrestrial equipment. "In the meantime, though, young Hanuman Saini, who learned to use soap and a towel, is going to have to get used to school and evenings without television. He isn't happy about it. 'Now I have no excuse for my parents when they want me to work,' he said."

Trehan, three years later, asked: "What are some of the long term effects?" A number of people felt that their knowledge about agriculture, health, and education had improved significantly. People also felt that the distinction among castes, which had been very sharp up to that point, had begun to soften, allowing them a little more freedom. A majority—85 percent—of the people in the non-tv village felt that television is needed for educational and informational purposes; only 68 percent in the tv-village felt so.

"Television, after its withdrawal, became a topic of discussion and concern for most people in the non-tv village. These people did not revolt against the administration. They only partially understood the final television broadcast concerning the withdrawal of television from their villages." Fewer than 25 percent asked anyone about it but, Trehan said, "because of the mistrust of district authorities, most villagers refrained from going to them."

About half the respondents in the non-tv village were either confused or did not know whom to blame for withdrawal of television. Only about 40 percent blamed the Indian and U.S. governments, and although the impression that the U.S. was responsible dropped to a very low level after two years, the feeling of distrust for local officials remained.

In the non-tv village, Trehan found "a strong feeling that television is for backward classes. Some upper-class people disliked SITE because it spread the 20 Points program, which included the message of nonpayment of loans, and abolition of bonded labor, and thereby reduced their apparent control in the village." Most respondents there switched to newspapers, after the withdrawal, which is consistent with their higher overall literacy; the reported rate of watching films also increased, especially among young men. "People switch over to the most convenient medium, as soon as one medium is disrupted," Trehan realized.

Over and above his social science research, Trehan collected some poignant quotes.

> A large number of respondents in both villages feel that a process of change has begun, . . . and essentially it is for a better life. One old man summed up his views of the changes in just two words, "Jehaniyat" and "Tamiz," meaning "worldwide knowledge" and "manners." When asked how he came to that conclusion, he said that it was written in the behavior of the people, who do not run away from investigators who visit their village now. Instead, they can communicate with—and even begin to trust—these strangers.
>
> The non-tv village respondants are even more dramatic in their assessment of television, its impact and its withdrawal. "We lost the light, and our hearts do not beat faster," they said. "We are silent."

AFTERWORD

Satellites may still seem esoteric because they are far away and expensive, but at least they are tangible things, made of metal and silicon, and placed in orbit by rockets and shuttles. These we understand because we understand the physical sciences that help to build and maintain them. The impact satellites have on law, politics, social groups, economics, and other human activities is also covered, to some extent, by the social sciences.

WHO OWNS THE RESOURCES?

The author believes that the central metaphor for communication is that it is a *resource,* that the key resource for satellite communication is the geostationary orbit, and that no matter whether there is a scarcity or an abundance of orbit slots or transponders, the resource should be managed with a concern for humanity, as one should manage any finite resource.

But there is a more subtle resource involved, which this book has only briefly covered. It is the radio spectrum—the whole range of frequencies over which electronic communication can take place. Seen in that light, satellites are just tools for obtaining and utilizing something more inherently precious. As mirrors and amplifiers for radio signals, satellites are comparable to mine shafts and smelters in the production of copper. Those must be built and operated efficiently, of course, but they serve mainly to obtain and develop the fundamental resource from the earth. The radio-frequency spectrum is the lode from which telecommunication can be mined.

Literally *global* controversies are raging over who shall be permitted to use what radio frequencies and to what purposes. Like oil, minerals, land, or water, the spectrum is a natural resource which has to be managed and conserved even while it is intensively used, bought, and sold. Understanding the importance of the radio spectrum is also fundamental to the communication revolution that is upon us, but it is a large subject in its own right, so it is covered in greater detail by the companion volume to this book, titled *Who Owns the Rainbow?*

APPENDIX

The sources for this book are varied. Some are direct quotations from published material; others are remarks or statements made in public presentations, such as conferences. In many cases, I also interviewed these people personally, or asked them to expand on their public pronouncements or published works.

As a result, I have taken some liberties with the use of quotation marks in *The Birds of Babel.* Where a quotation is attributed directly to a person, it fairly represents that person's point of view, in his or her own words. Those words, however, may have been expressed at different times, or they may be condensed from published papers, where a common point is made in otherwise separate paragraphs. I have done this only for clarity and directness. I wanted to avoid the use of elipses or brackets, whenever possible, because I consider them distracting, and more appropriate in textbooks.

Many interviews were conducted at national or international conferences, over several years. They enabled me to contact people more efficiently than by trekking to their places of business.

The conference whose published proceedings I have relied on most often is the annual Pacific Telecommunications Conference, sponsored since 1979 by the Pacific Telecommunications Council. Copies—if not available in school libraries—may be purchased from the council: 1110 University Avenue, Suite 303; Honolulu, HI 96826. The citations for proceedings are given as "Proc.PTC," followed by the year of the conference; oral presentations, such as participation in a roundtable discussion, or personal interviews during the conference are given as "Pres.PTC" (presentation at PTC) or as "Int.PTC" (interviewed at PTC).

Two other conferences did not publish proceedings. The Satellite Communications Users' Conference ("SCUC") sells audio tapes of its sessions. Contact the sponsor: Cardiff Publishing Co. 6430 South Yosemite St., Englewood, CO 80111. The Satellite Summit ("Summit") was sponsored by Satellite Week (news service), published by Television Digest, Inc. 1836 Jefferson St., Washington, DC 20036.

Where an interview was conducted privately, I have noted that as "Int." and the year it was done.

Books are cited in the style used by the social sciences. Articles are cited with headlines or titles where possible, to suggest the flavor of the piece. The most-used sources are *Satellite Week* news service ("SatWeek"), *Satellite Communications* magazine ("SatComm"), and *The Wall Street Journal* ("WSJ"), which devotes a great deal of coverage to high-technology in general, and data processing, satellites, and telecommunications in particular.

Names of other periodicals are given in full.

AFTER COMPUTERS . . . COMMUNICATION

Martin, James *Future Developments in Telecommunication* (2nd Edition) Prentice-Hall, Englewood Cliffs, NJ 1977

Vanguard and the American satellite programs prior to Sputnik: "A Scientific Race Against Time to Launch the First Man-Made Moon" Life (Time-Life, Inc.) June 3, 1957

Barnouw, Erik *A History of Broadcasting in The United States* Vol. 1 *A Tower In Babel* Oxford University Press 1966

Einstein's remarks: from an address to the California Institute of Technology, 1931 (ref: Bartlett's Familiar Quotations)

THE GEOSTATIONARY ORBIT

Clarke's ideas about the geostationary orbit appeared in the magazine *Wireless World* in 1946, but have been accurately paraphrased in practically every book about satellites ever since.

"RCA's Lost Satellite Would Seem to Be Found" WSJ March 16, 1983

Space stations: ref. Smith, Delbert D. and Rothblatt, Martin A. "Geostationary Platforms: Legal Estates in Space" *Journal of Space Law:* Vol. 10, Number 1, Spring 1982

TRADE-OFFS IN SATELLITE COMMUNICATION

Much of the technical material on trade-offs in telecommunications generally, and on FDM and TDMA specifically, comes from Martin, James

Future Developments in Telecommunications (2nd Edition) Prentice-Hall, Englewood Cliffs, NJ 1977

A good, nontechnical description of how satellites work is in Easton, Anthony T. *The Home Satellite TV Book* Wideview Books, San Francisco, CA 1982

Reports on the use of the U.S. Space Shuttle to launch communication satellites were carried by United Press International November 12, 1982, and by WSJ November 5, 9, 12, and 17, 1982. Additional information appeared in an advertisement in WSJ November 18, 1982, paid for by Rockwell International (which builds the shuttle).

"Conestoga" rocket background from SSI information kit, September 1982

Telstar and Comstar versions of Hughes HB-376 explained by Morgan, Walter (columnist) SatComm December 1982

Four DBS slot allocations: Reinhart, Edward E. "Three Is Not Enough" SatComm December 1982

Port Authority quoted and Teleport described in WSJ June 11, 1982 and in Information Systems News, CMP Publications, Manhasset, NY June, 1982

SBS experience with rain: Hall, Robert—Summit 1982

PARALLELS OF HISTORY

Much of this chapter is based on facts accumulated by Barnouw, *op-.cit.,* and from vignettes in Howat, Bruce B., *Great Moments in Communication* Communications News, Geneva, IL 1973

ITU history from Codding, George A. and Rutkowski, Anthony M., *The ITU in a Changing World* Artech House, Dedham, MA 1982

"Shreveport Doctrine" from Ahern, Veronica—Int. 1982

"Public interest, convenience and necessity" from Communications Act of 1934; complete text in Title 47 United States Code

Transatlantic cables in WWII from Sampson, Anthony *The Sovereign State of ITT* Fawcett Publications, Greenwich, CT 1974

Baxter's remarks made at the National Telecommunications Association December 3, 1982; quoted in Communications Daily (Television Digest, Inc., Washington, DC) December 6, 1982

SPACE LAW

Hosenball, Neil—Int. 1982

Ploman, Edward W. "Satellite Broadcasting: A New Era of International Law" in Richstad, Jim ed. *New Perspectives in International Communication: A Report on Fair Communication Policy for the International Exchange of Information* East-West Communication Institute, Honolulu, HI 1977. Also Int. PTC'83

Background material on international legal issues (*a priori* allocation, direct-broadcast satellite TV, etc.) from Richstad, Jim and Harms, L.S. eds, *Evolving Perspectives on the Right To Communicate* East-West Communication Institute, Honolulu, HI August, 1977

The idea of communication as a human right was articulated in the 1948 United Nations Declaration of Human Rights, Article 19

Borgese, Elizabeth Mann "The Law of the Sea" in Scientific American March, 1983

Butler, Richard—Int.PTC'83

"New actors" from Ploman, Edward W., unpublished paper Pres.PTC'83 and Int.PTC'83

Ashbacker-type proceedings and FCC rules on CPE: Ahern, Veronica—Int.1982

Melody, William—Pres.PTC'82 and Int.PTC'83

Jussawalla, Meheroo—Pres.PTC'83 and Int.PTC'83

Butler, Richard—Int.PTC'83

Levin, Harvey J. "Alternative Strategies for Managing the Orbit and Spectrum Resource" Proc.PTC'81. Also Int.PTC'83

Pelton, Joseph—Pres.PTC'83 and Int.PTC'83

Hudson, Heather—Pres.PTC'83 and Int.PTC'83

PIRACY ON THE HIGH FRONTIER

McGeer, Patrick—Int.1981

Belanger, Charles—Int.1981

Jacobsen, Dennis—Int.1981

Israel, Mike—Int.1983

INTELSAT

Masuda, Motoichi; Alessandrini, Adolfo; and Alegrett, Jose—quoted in *INTELSAT Memoirs* (INTELSAT 15th anniversary publication) August 20, 1979

Additional reference material on INTELSAT from:

Agreements Relating to the International Telecommunications Satellite Organization "INTELSAT"—August 20, 1971; entered into force February 12, 1973.

Handbook on INTELSAT (published by INTELSAT) May 1, 1982

INTELSAT Research and Development Program: 10 Years of Progress (published by INTELSAT) 1981

Pelton, Joseph; Peres, M.; and Ashok, S. *INTELSAT: The Global Telecommunications Network*—Pres.PTC'83

Pirard, Theo "Intersputnik: The Eastern 'Brother' of INTELSAT" Sat-Comm August, 1982

Long, Mark and Keating, Jeffrey *The World of Satellite TV* The Book Publishing Co. Summertown, PA 1983

Caruso, Andrea—Summit 1982

Newman, Barry "European States Face Problem of Controlling Their Neighbors' TV"—WSJ March 23, 1982

INMARSAT

Background on INMARSAT from *Ocean Voice:* The Journal of Maritime Satellite Communications (published by INMARSAT)

Lundbert, Olaf—Summit 1982

Davis, Sir Thomas—(keynote speaker) Pres.PTC'82

Perras, Marcel—Pres.PTC'82

Lovell, Robert "The Status of NASA's Communications Program"—Pres.PTC'82

Hudson, Heather—Pres.PTC'82

"Satellite Communications for the Pacific Islands" Final Project Report Public Service Satellite Consortium, Washington, DC December, 1982—Pres.PTC'83

"A Study on the Pacific Regional Satellite Communications System" Research Institute of Telecommunications and Economics Tokyo, Japan January, 1983—Pres.PTC'83

(GTE position) Vallo, Roger P. "The New Economics of Pacific Telecommunications"—Proc.PTC'81

Pelton, Joseph—Pres.PTC'83

ATS-1: PROMISES AND COMPROMISES

Much of the background on ATS-1 and the PEACESAT project stems from interviews with the PEACESAT project director John Bystrom and the terminal manager Carol Misko, in Honolulu, during 1977–79. Also useful were reports published by PEACESAT, including *PEACESAT Project Early Experience* Report One: October, 1975.

Papua New Guinea experience—Proc.PTC'81

Porter, Kit—Pres.PTC'81

Dator, Jim "EIES and Racter and Me: Computer Conferencing from a Pacific Island"—Proc.PTC'80

University of the South Pacific experience—Pres.PTC'82

Southworth, John—Int.PTC'82

Kingan, Stuart G. "PEACESAT Overcame the Cook Islands' Problem of Isolation" *The Printout* (magazine) Honolulu, HI March, 1980; adapted from Pres.PTC'80

Plant, Christopher M. "Analysis Reveals PEACESAT Is Dominated By Metropolitan Terminals" *The Printout* (magazine) Honolulu, HI March, 1980; adapted from Pres.PTC'80

Bystrom, John at 1981 PEACESAT Users' Meeting in Wellington, N.Z. quoted in Flavell, Elsa and Bystrom, John "A Discussion at Wellington on Satellite Communication" *Pacific Islands Communication Journal* Vol. 11 No. 1 1982 (published by Pacific Islands News Association Suva, Fiji)

Ritter, Craig "Teleconferencing Networks for Developing Societies: A Brief Analysis"—Proc.PTC'81

Shaginaw, George "Alaska and ALASCOM: A Decade of Telecommunications Development"—Proc.PTC'82

Plummer, Sioux "Alaska's Legislative Teleconference Network—Why It Works"—Proc.PTC'81

SATELLITES FOR THE PACIFIC

Background for this chapter comes from the proceedings of the PTC for the years 1979–83

Ippolito, Louis—Pres.PTC'83

Canada-U.S. agreement to share satellite network—WSJ August 27, 1982

Halprin, P.W.—Int.1981

Adams, Barry—Int.1981

Stewart, Robert—Int.1981

Israel, Mike—Int.PTC'83

VERY HIGH FINANCE

Stine, G. Harry—Sat.Comm. February, 1980

Some background on rate-setting and the role of economics in telecommunication comes from Dordick, Herbert—Pres.PTC'81, Pres.PTC'82 and subsequent interviews.

Tariffs quoted from American Satellite Co., STARNET Corp., and Bonneville Satellite Corp. were current in August, 1982

Melody, William "The Economics of Information as Resource and Product"—Proc.PTC'81

Survey of Asian professionals in Randolph, Robert H. "Computer Networks for Asia-Pacific Economy"—Proc. PTC'81

FINANCING SATELLITES

Much of the background for this chapter comes from Pafumi, Glen—SCUC 1982 and Int.1982

Comsat de jure monopoly: Communications Satellite Act of 1962: Title 47 United States Code

Miller, Jonathan—Int.1982

Hughes, Brian—Sat.Comm. August, 1981

New local phone rates: WSJ March 28, 1983

TRANSPONDERS FOR SALE: REAL PROPERTY IN OUTER SPACE

Besides the sources listed below, additional information about transponder sales comes from Whitehead, Clay T.—Summit 1982; Pafumi, Glen—Int.1982; Will, Tom—Int.1982; and Williams, David—Int.1983

Miller, Jonathan—Int.1982

FCC Docket No. 82-45 In the Matter of Domestic Fixed-Satellite Transponder Sales: Memorandum Opinion, Order and Authorization Adopted July 29, 1982; Released August 17, 1982 (decision favoring transponder sales; includes dissent by Commissioner Fogarty)

Winnik, Joel—SCUC 1982

Wold, Robert—SCUC 1982; Int.PTC'80

Johnson, Jack—SCUC 1982

Clifford, Gregory—SCUC 1982

Kaseman, Carl—SCUC 1982

Paschall, Lee—Summit 1982

Auction and lottery sales invalidated—WSJ January 29, 1982; February 9, 1982

Transponder Advertisement in WSJ January 18, 1982

Attorney who placed a transponder ad, quoted in Sat.Week December, 6, 1982

Cooney, John "Lowering Skies for the Satellite Business" *Fortune* December 13, 1982

DIRECT-TO-HOME BROADCAST SATELLITES (DBS)

Background material on DBS comes from Sat.Week, Sat.Comm. and Satellite News during 1981–83. Additional material from promotional literature distributed by prospective DBS operators and carriers. Technical background from Prichard, Wilbur L. and Kase, Charles A. "Getting Set for Direct-Broadcast Satellites" IEEE Spectrum August, 1981

Easton, Anthony T. *The Home Satellite TV Book* Wideview Books San Francisco, CA 1982

Reinhart, Edward—SCUC 1982

PICTURES FROM SPACE

Reports on Cosmos 954 from United Press International and Associated Press January, 1978

O'Toole, T. and Babcock, C. "How the CIA 'Lost' Spy Data to Soviets" Washington Post Service, from Honolulu *Advertiser* August 23, 1978

Oatis, W.N. "U.S. Admits One Sat In Worldwide Nuclear Spill" Associated Press, from Honolulu *Star-Bulletin* March 28, 1978

"The Real War In Space" (television documentary) produced by WGBH Boston, for NOVA (distributed by PBS) 1980

LANDSAT background material supplied by Earth Resources Orbiting Satellite (EROS) Data Center, Sioux Falls, SD

State of Hawaii LANDSAT use—Int.1979

"Should We Industrialize Space?" NASA Tech Brief (MFS-23963) Vol. 5 No. 1 Spring, 1980

Large, Arlen "Reagan Wants to Sell Weather Sats to Highest Bidder Among U.S. Businesses"—WSJ March 9, 1983

GOES-4 failure—Sat.Week December 6, 1982

Brandli, Henry W., Lt. Col. USAF (Ret.) "The Coming of Age in Military Meteorology" *Sat.Comm.* June, 1982

POSITION-LOCATING SATELLITES

Much of the background for this chapter comes from interviews with Stephen G. Glatzer, 1982–83

SARSAT System Summary April, 1980 and *SARSAT* 1982 (published by NASA Goddard Space Flight Center)

INMARSAT geopositioning system background from Lundberg, Olaf—Summit 1982

O'Neill, Gerard K. and Schmidt, Charles Int. Summit 1982 and "Satellites That Fight Muggers" *New York Times* May 12, 1983

DON'T COUNT CABLES OUT

Dawidziuk, Bogumil M. "Undersea Systems Technology Trends"—Proc.PTC'82 and Int.PTC'82

Fournier, Martin and Pal, Nirmal E. "Development of the ANZCAN Project"—Proc.PTC'82

Lake Washington fiber-optic cable—Int.1981

WE ARE SILENT

Trehan, Vijay R. "We Are Silent: Effects of Withdrawal of Television from Satellite Instructional Television Experiment (SITE) Villages in India"—Proc.PTC'80 and Int.PTC'80

Chutkow, Paul "TV 'School' Ends in India; Results Live On" Associated Press, from Honolulu *Star-Bulletin* August 1, 1976

ADDITIONAL MATERIAL

Besides the sources listed in this Appendix, the reader may find additional information in the following periodicals:

Satellite News Phillips Publishing Co. Bethesda, MD

Telephony Telephony Publishing Corp. Chicago, IL

Telephone Engineer and Management Harcourt Brace Jovanovich Publications, Geneva, IL

SAT-Guide Commtek Publishing Co. Hailey, ID

Satellite TV Wiesner Publishing Co. Littleton, CO

Teleconference VSN Satellite Communications Services San Ramon, CA

Satellite TV Week Fortuna Publishing Co. Fortuna, CA

INDEX